# Life Cycle Assessment for Sustainable Mining

# Life Cycle Assessment for Sustainable Mining

**Dr. Shahjadi Hisan Farjana**
Department of Mechanical Engineering, University of Melbourne, Melbourne, VIC, Australia

**Dr. M. A. Parvez Mahmud**
School of Engineering, Deakin University, Geelong, VIC, Australia

**Dr. Nazmul Huda**
School of Engineering, Macquarie University, Sydney, NSW, Australia

ELSEVIER

Elsevier
Radarweg 29, PO Box 211, 1000 AE Amsterdam, Netherlands
The Boulevard, Langford Lane, Kidlington, Oxford OX5 1GB, United Kingdom
50 Hampshire Street, 5th Floor, Cambridge, MA 02139, United States

**Notices**
Knowledge and best practice in this field are constantly changing. As new research and experience broaden our understanding, changes in research methods, professional practices, or medical treatment may become necessary.

Practitioners and researchers must always rely on their own experience and knowledge in evaluating and using any information, methods, compounds, or experiments described herein. In using such information or methods they should be mindful of their own safety and the safety of others, including parties for whom they have a professional responsibility.

To the fullest extent of the law, neither the Publisher nor the authors, contributors, or editors, assume any liability for any injury and/or damage to persons or property as a matter of products liability, negligence or otherwise, or from any use or operation of any methods, products, instructions, or ideas contained in the material herein.

**Library of Congress Cataloging-in-Publication Data**
A catalog record for this book is available from the Library of Congress

**British Library Cataloguing-in-Publication Data**
A catalogue record for this book is available from the British Library

ISBN: 978-0-323-85451-1

For information on all Elsevier publications visit our website at
https://www.elsevier.com/books-and-journals

*Publisher:* Candice Janco
*Acquisitions Editor:* Marisa LaFleur
*Editorial Project Manager:* Aleksandra Packowska
*Production Project Manager:* Debasish Ghosh
*Cover Designer:* Victoria Pearson-Esser

Typeset by TNQ Technologies

Working together to grow libraries in developing countries
www.elsevier.com • www.bookaid.org

# Contents

# Preface

The term 'sustainable mining' refers to the employment of technologies and best practices to reduce the environmental impacts associated with the mining, extraction and processing of minerals. Generally, environmental impacts caused by mining involves soil erosion, acid mining drainage, contamination of water resources, the formation of sinkholes, affecting human health through carcinogenic and noncarcinogenic substances. To avoid these detrimental effects, mining companies should strictly adhere to the environmental regulations and codes to attain government policy. However, these impacts can significantly be reduced by appropriately identifying and taking measures to reduce them. Life cycle assessment (LCA) is a powerful tool to quantify environmental impacts for hotspot identification to promote sustainable production system design. The applicability of the knowledge of LCA is verified through this book, which would be extremely beneficial for mining or manufacturing engineering students or graduates. Chapter 1 provides the basics of LCA. Chapter 2 systematically presents the survey of existing literatures on LCA of mining industries in respect of sustainability. Chapter 3 presents the case study of the life cycle inventory development, systems modelling and analysis of ilmenite and rutile mining processes in Australia. Chapter 4 presents comparative life cycle impact analysis including the material flow analysis starting from the modelling to results interpretation for three different types of uranium extraction processes. Chapter 5 shows how to conduct the LCA of the beneficiation process of gold—silver—lead—zinc—copper combined production process. Chapter 6 shows the analysis of the solar process heat integration feasibility in mining industries based on LCA. This book is originated from the PhD thesis conducted by the leading author Dr Shahjadi Hisan Farjana done at Macquarie University, Sydney, Australia. Special thanks to the coauthors of this book and Elsevier for publishing this book. We would also extend our thanks to our families for their support.

**Dr Shahjadi Hisan Farjana**
**Dr M. A. Parvez Mahmud**
**Dr Nazmul Huda**

# List of Abbreviations and Symbols

| | |
|---|---|
| AP | Acidification |
| Bq C-14 eq. | Bq C-14 equivalents into the air for ionising radiation |
| $C_2H_3Cl$ eq | kg chloroethylene equivalents into the air, for carcinogens and noncarcinogens |
| CC | Climate change |
| CED | Cumulative energy demand |
| CETEM | Center for mineral technology database |
| CML | Center for methodological development |
| CSIRO | Commonwealth Scientific and Industrial Research Organization |
| CST | Concentrated solar thermal technology |
| CTUe | Comparative toxic unit for ecosystems |
| CTUh | Comparative toxic unit for human health |
| DALY | Disability-adjusted life year |
| DNi | Direct nickel method |
| EDIP | Environmental Design of Industrial Products |
| ETC | Evacuated tube collector |
| EU | Eutrophication |
| FEU | Freshwater eutrophication |
| FFD | Fossil fuel depletion |
| FPC | Flat plate collector |
| FWE | Freshwater ecotoxicity |
| GER | Gross energy requirements |
| GWP | Global warming potential |
| HH | Human health |
| HPAL | High-pressure acid leaching |
| HT | Human toxicity |
| IAI | International Aluminum Institute |
| ILCD | International Reference Life Cycle Data System |
| IPCC | Intergovernmental Panel on Climate Change |
| ISO | International Organization for Standardization |
| kBq U235 eq | A decay of 1000 U 235 nuclei per second |
| kg C deficit | Kilograms of carbon deficit |
| Kg $C_2H_4$ eq | kg ethylene equivalents into the air for respiratory organics |
| kg CFC-11 eq | Ozone depletion potential OZDP kg CFC-11 eq |
| kg $CO_2$ eq | Carbon dioxide equivalent |
| kg N eq | Eutrophication potential for air emissions |

| | |
|---|---|
| **kg NMVOC eq** | Nonmethane volatile organic compounds (NMVOCs) equivalent units |
| **kg O₃ eq.** | A kilogram of ozone equivalent |
| **kg P eq** | Freshwater eutrophication kg P eq |
| **kg PM2.5 eq** | Human Health Particulate |
| **Kg PO₄ eq** | kg $PO_4$ − equivalents into a P-limited water aquatic eutrophication |
| **kg Sb eq** | Abiotic depletion equivalent |
| **Kg SO₂ eq** | kg $SO_2$ equivalents into the air for acidification |
| **Kg TEG water or soil** | kg triethylene glycol equivalents into the water for aquatic ecotoxicity and soil for terrestrial ecotoxicity |
| **LCA** | Life cycle assessment |
| **LCA-Pro** | Life cycle assessment software name |
| **LCI** | Life cycle inventory |
| **M²a** | Metre square times year |
| **M²org.arable** | $m^2$ organic arable land for land occupation |
| **M³ H₂O** | The volume of water supply |
| **m³ water eq** | Volume of water |
| **ME** | Marine eutrophication |
| **MJ HHV** | Higher heating value in megajoule |
| **MJ primary** | Total life cycle primary energy use |
| **MJ primary nonrenewable** | MJ primary nonrenewable for nonrenewable energy |
| **MJ surplus** | Characterised fossil fuel profile |
| **molc H+ eq** | Acidification units |
| **molc N eq** | Terrestrial eutrophication |
| **MT** | Mega tonne |
| **Non-CST** | Nonconcentrated solar thermal technology |
| **ODP** | Ozone depletion potential |
| **PDF*m²*yr** | Potentially disappeared fraction of species over a certain area over a certain time |
| **PMF** | Particulate matter formation |
| **POCP** | Photo-oxidant creation potential |
| **TAP** | Terrestrial acidification |
| **Term** | Description |
| **USGS** | US Geological Survey database |
| **WD** | Water resource depletion |
| **WMO** | World Meteorological Organization |
| **WSP** | Water scarcity potential |
| **μPt** | Micro points |

Chapter 1

# Introduction to Life Cycle Assessment

## Definition of Life Cycle Assessment

Life cycle thinking is the way of thinking of the consequences in the environmental, economic and social effects of a product throughout its entire life. Life cycle assessment (LCA) is the steady-state, global/regional, comprehensive and quantitative analysis of environmental or social impacts of a product/process/system of processes from its entire life cycle from beginning to end — which means the effects on ecology, resources and human health. The life cycle stages include all the raw material, resource and energy consumed through the manufacturing stages including the raw materials acquisition stage, processing stage, manufacturing stage, product life phase, and waste management/end-of-life scenario. At the same time, transportation is inclusive in every step. However, the inclusion of life cycle stages should be defined by the system boundary considered for a particular study. The system boundary can be cradle-to-gate, cradle-to-grave, gate-to-gate, or gate-to-grave. It might also be called life cycle analysis or life cycle thinking. The conceptual framework developed based on ISO 14040 to ISO 14044 helps the environmental management and technologists to meet the standards of sustainable development through life cycle assessment. Among the criteria of sustainable development, it requires substantial improvement on the eco-efficiency and reduced greenhouse gas emissions on human health, ecosystems and resources. Each manufacturers or suppliers is responsible for ensuring sustainability through product stewardship (ISO, 2004).

## Applications of LCA

LCA is a sustainable decision support tool for product/process improvement of a company. The development can be on design, manufacturing, use phase, or end-of-life phase of a product. To ensure sustainability throughout the entire supply chain, the upstream or downstream manufacturers should prove that their products meet the justified sustainability standards. From the stakeholder's perspective, it is an integral part of environmental management — not

Life Cycle Assessment for Sustainable Mining. https://doi.org/10.1016/B978-0-323-85451-1.00001-9

only for product development but also for developing the strategic policy of sustainable manufacturing. But for the quality data-LCA studies must ascertain the accuracy of the analysis. To educate engineers, environmental scientists and technologists and for raising the public awareness and knowledge of LCA studies requires professional training, peer review, public seminars and workshops, blockchain-based processing of inventory datasets, assurance of credible datasets and stakeholder engagement. A successful LCA study required for the corporate sustainability reporting of a manufacturing company involves comparable conceptual framework, indicators and benchmarks. Companies can choose their metrics and impact assessment indicators, which makes it difficult for stakeholders to compare one LCA study with another.

There is also a lack of data quality checking metrics in LCA studies which raise the question towards credibility and data accuracy. The benchmarking for credible comparison of LCA studies within the same domain is essential if the LCA study would be published publicly.

For accuracy and enhanced data quality, credibility for analysis and benchmarking for comparison, the standard framework for LCA study for an industrial domain can be developed through explicit instructions for system boundary development, environmental metrics and indicators, quality assurance of transparent data for the construction of life cycle inventory datasets, the inclusion of uncertainties while compiling datasets for inventory database.

The main applications of LCA internally or externally for a manufacturing company are:

- Internal industrial use of product or process development.
- Strategic planning and decision support for the internal use of industries.
- Reduce the costs of production.
- Minimising the damage to the environment and human health.
- External use for marketing through the achievement of sustainable development goals.
- Comparison of different products or manufacturing systems within the same domain of industry, system boundary and functional unit, same systematic framework.
- Public policy generation through sustainable development goals.

## Use of Environmental Information from LCA in Decision-making

- For planning and capital investment in green design and waste management.
- Eco-design and product development.
- Green procurement or operational management.
- Ecolabelling for communication and marketing-verified certification of environmental labelling using multicriteria or predefined set of criteria.

- Financial management through cutting carbon taxes.
- Environmental emission regulations.
- Life cycle thinking/life cycle management.
- Design for environment.
- Cleaner technology development.

## Levels of LCA

LCA methodology can be categorised into three levels based on technological details:

- **Conceptual LCA** − First level of LCA based on limited environmental aspects of few life cycle stages where there is still some improvement potential existing for the manufacturer. The results might be useful for qualitative reporting of assessment results, but not suitable for corporate marketing or explicit publication of LCA study.
- **Simplified LCA** − This is the type of comprehensive assessment using generic datasets covering the whole life cycle of a product/system of processes. The time required and expenditures as well reduce significantly here, which is a significant difference from detailed LCA. This consists of a screening of life cycle stages, simplification of LCA results for future recommendation and assuring the reliability of the analysis results. This is often termed as 'Streamlined LCA'.
- **Detailed LCA** − This type of LCA is comprehensive with the full consideration of each life cycle stages with system-specific datasets and analysed in detail for further process improvement.

## Essential Steps of Life Cycle Assessment

The significant steps of an LCA study consist of four essential stages based on ISO 14040(Environmental management − Life cycle assessment − Principles and framework). Table 1 describes the major steps to be covered during an LCA study.

## Goal and Scope Definition

Based on ISO 14041: Environmental management − Life cycle assessment − Goal and scope definition and inventory analysis. The main issues to be addressed in this phase are goal and scope definition (ISO, 2004).

    **Goal** − The purpose of conducting the LCA study while also mentioning the audience of the results produced. It can be the comparison of different products with the same functional unit, same purpose/use of those products. It can also be defined as the improvement potential of product/process through

**TABLE 1** Life cycle assessment (LCA) systematic procedure.

| Stages | Major components and brief description |
| --- | --- |
| Goal and scope definition | Goal — the scope of the study should be described, followed by the reason to carry out and intended use. The list of audience should be provided. The public release capability of data might be justified |
| | Scope — inclusions of the LCA analysis. |
| | System boundary — processes considered within the system. |
| | Function — use of the output product. |
| | Functional unit — per unit of the function of the product. |
| | Reference flow — the unit amount of each inventory to fulfil the function of that functional unit. |
| Life cycle inventory analysis | Data collection — inventory datasets collected for the system of processes within system boundary, based on time, geography and technological coverage. |
| | Data validation per unit process — mass or energy balance of the collected datasets against the unit process. |
| | Data validation per functional unit — calculation of an environmental load of each element per functional unit. |
| | Data aggregation — same product originated from different processes are aggregated for simplification of the presentation of inventory. |
| | Refining the system boundary — after the balancing of datasets, the system boundary might require to be improved, which is not mandatory. |
| | Allocation — partitioning of the product flows based on mass/economic value of the by-products/coproducts and required for multiple input or multiple output process. It can be avoided by dividing the unit process or by expanding the system boundary. Allocation is also required in case of waste recycling — based on closed-loop approximation, avoided impact approach and cut-off method. |
| Life cycle impact assessment | Classification — classification of environmental loads based on impact categories. The impact categories are discussed separately in the next section. |
| | Characterisation — relative contribution of an inventory parameter for a particular impact category. After calculating for each parameter, the total contribution of inventories per impact category is aggregated. |
| | Normalisation — valuation of the characterised impact of an inventory element within one impact category. |
| | Weighting — imposing relative significance among the impact categories. |

**TABLE 1** Life cycle assessment (LCA) systematic procedure.—cont'd

| Stages | Major components and brief description |
| --- | --- |
| Results interpretation | Key issues identification − identification of the product/process/element which is responsible for high impacts on the environment. |
| | Checking for completeness, sensitivity, consistency and data quality − the completeness should be checked based on data availability and individual results. In the case of missing datasets, goal and scope of the study must be revised. Sensitivity and uncertainty analysis can be conducted for assumptions based on process optimisation and identification of hotspots. Consistency of overall calculation and assumptions for system boundary, temporal coverage, allocation and impact calculation should be justified. |
| | Sensitivity analysis − sensitivity analysis is finding the sensitive parameters or choices within the system of processes which can be replaced by environmentally sensible choices or reduced in content to decrease the environmental burdens. |

innovation and hotspot analysis. It describes what is going to be reported at the end of the analysis.

In summary, the goal will justify what type of LCA would be performed: descriptive LCA (based on definitions and theories) or change-based LCA (based on empirical mode)

**Scope** − The inclusion and exclusion in the focused LCA study about the process and also about the methodology. The function of the system or the unit of function of the product analysed should be clarified in this stage. The inventory datasets collected should be quantified according to the functional unit. This is the reference according to which the collected datasets are normalised. System boundary or the inclusion of the life cycle stages within the assessment conducted should be justified; it could be the input and output elements within the boundary of a process or a system of processes. Inclusion of impact categories what would be the coverage of the study is also clarified at this stage of LCA. Data quality, requirements, assumptions and limitations with the allocation procedure between multiple products should be described (European Commission - Joint Research Centre - Institute for Environment and Sustainability, 2010b).

Most importantly, any omission of a particular production process should be clearly described with justified reason. To fulfil the function, the required amount of product is the reference flow. There are some different types of flow: elementary flow, product flow and intermediate flow. Elementary flow enters and leaves the product system without human transformation, including material/energy. Product flow is the product entering or leaving the product system. Intermediate flows are flows occurring between unit processes within the system.

The cut-off criteria can be utilised in a study which is the specification of the least amount of environmentally significant material/energy flow to be considered in a study. If it is insignificant, the particular flow can be excluded and mentioned in the scope section of the survey. The summary of the scope of an LCA study would comprise coverage of temporal conditions, geography, technology, processes, interventions and impacts (Braune and Duran, 2018).

## Life Cycle Inventory Analysis

This phase includes the datasets compiled based on input materials, energy and resources, and output products and emissions towards the environment (ISO 14041: Environmental management − Life cycle assessment − Goal and scope definition and inventory analysis). The data quality should be analysed based on coverage of time length or validity, specification based on manufacturing/process geographic location and type of the technology mix. The indicators of the quality are accuracy, reliability, consistency and representativeness (Frischknecht et al., 2007).

Generally, this phase consists of the data collection, refining the system boundary, data calculation and validation, and allocation. The datasets related to the material consumption, wastes, outputs and emissions are collected/calculated, which can be site-specific or averaged for different sites located in the same or different geographic region. The inventory datasets can be qualitative or quantitative. Quantitative data are more useful for comparing materials/processes but is time dependent for data validity and sometimes less accurate. Qualitative data for single operations are as well valid for longer term, which is used for descriptive analysis. The significant steps of data validation include the data quality check and accuracy measurement based on the mass or energy balance and comparative emission factor analysis. The nature of data validation is iterative. If the data validation results with inaccuracy, the missing data can be replaced based on justified sourcing of datasets from secondary sources. The recalculated value can be an acceptable value reported in literature or database or a zero value, or a calculated/adjusted/averaged value (Lee and Inaba, 2004).

The datasets can be from:

- Primary data − Direct measurement, models or samples, data provided by manufacturers and product or company report.
- Secondary data − Averaged data, data from the LCA database and data from literature resources (Scientific journals or reports).

The types of data required for LCA are:

- Inputs − Energy (electricity, process, transportation, combustion, renewable and nonrenewable electricity), water inputs, gas and oil, chemicals and capital equipment.

- Outputs − Product/coproducts/by-products. Atmospheric emissions like particulates, nitrogen oxides, organic compounds, sulphur oxides, aldehydes, ammonia, lead and carbon monoxides. Waterborne emissions like biological oxygen demands, suspended and dissolved solids, oil, sulphides, grease, chromium, fluorides, phosphates and ammonia. Wastes like industrial waste generated during production, process waste which is not recycled, wastes from fuel consumption and wastes from products and packaging.

The standard formats for LCA data are:

- ISO 14048.
- Spold/EcoSpold.
- Spine.
- UNEP/SETAC.
- ELCD.

For the multioutput process/processes with open-loop recycling, allocation based on mass/economy/geography should be adopted into the entire life cycle inventory datasets − materials, resources and emissions. If it is not possible to include all the inputs, outputs and emissions within the system boundary, then allocation can be avoided. In this case, it is considered that no additional outputs are flowing outside the system boundary. If the unit process can be divided, then divide the unit process into subprocesses and avoid allocation. Otherwise, expand the product system and then prevent the distribution. If allocation cannot be avoided, while allocation is under consideration, then all the inputs and output emissions should be partitioned among the coproducts. If it is possible to divide the inventory materials physically, then the allocation is reasonable by sharing them physically. Otherwise, the distribution must be done using any other forms (Zbicinski and Stavenuiter, 2006).

Allocation for end-of-life scenario can be of two types: open loop and closed loop. For closed-loop recycling, allocation is relatively straightforward, whereas open-loop recycling method- allocation can be of using end-of-life method, recycled content method, equal parts, or mutually dividing the specific recycling processes among systems (Weidema, 2012).

## Life Cycle Impact Assessment

This step conducts the impact analysis of the manufacturing process or system of operations in the quantitative form of emissions (ISO 14042: Environmental management − Life cycle assessment − Life cycle impact assessment). It is the third phase where impact categories to be assessed are determined, followed by the classification, characterisation, normalisation, and weighting. The environmental impact categories can be: abiotic resources depletion, land use, global warming potential (GWP), ozone layer depletion, ecotoxicity, human toxicity, photochemical ozone formation, acidification and eutrophication (Ciroth and Gmbh, 2016).

The first step is the categorisation of environmental impacts, which is a decision-making process of goal and scope definition of a particular LCA study. Not all the impact categories are equally crucial for a process or system of operations.

Classification is the allotment of life cycle inventory inputs and outputs to impact categories. This quantitative step involves the multiple mentioning of the same inventory if they contribute to numerous groups and do not affect each other through double counting (Wedema et al., 2013).

Characterisation is the basis of aggregation of inventory datasets for individual impact categories relating to the inputs and outputs. Based on ISO 14040, this is a factor derived from the characterisation model to convert the life cycle inventory results to the category indicator. The validation of characterisation is dependent on spatial and temporal coverage of the impact categories within the inventory. In simple words, it quantitatively shows how each inventory data contributes to each impact categories (European Commission - Joint Research Centre - Institute for Environment and Sustainability, 2010a).

For calculating the impact indicator, the formula is

Impact indicator = Inventory data × Characterisation factor.

Three main types of choice for impact indicator are:

- Midpoint-based indicator — Closer to interventions which rely on scientific information and proven facts while uncertainty is limited. For example, methods which follow midpoint-based signs are CML-IA and TRACI.
- Endpoint-based indicator — Closer to endpoints which relies on presenting information is the easily understandable manner to interpret. For example, method which follow endpoint-based signs are Eco-Indicator 99.
- Combination of midpoint- and endpoint-based manner — These methods combine two different impact indicator choices which provide the option to compare the midpoint- and endpoint-based results easily. For example, ways which follow both midpoint- and endpoint-based indicators are ReCiPe and Impact 2002+.

Based on ISO 14040, the impact categories are the class which represents the environmental issues concerning the life cycle inventory results to be assigned. The commonly used impact categories are:

- Global impacts — Global warming, ozone depletion, and resources depletion.
- Regional impacts — Photochemical smog and acidification.
- Local impacts — Human toxicity, terrestrial toxicity and aquatic toxicity.

Normalisation is the calculation of the magnitude of category indicator results to reference information per impact category. The function of normalisation is to understand the relative importance of each indicator results. The calculation procedure is as follows, while the unit is the year:

Normalised Indicator Result = Characterised value of impact indicator result per category/Normalised reference value per category.

Weighting is the qualitative or quantitative and comparative analysis of the impact categories based on the characterisation factors working on each of them. Based on ISO 14040, the weighting is the way of converting and aggregating the impact indicator results across the impact categories. The function of this step is to calculate the relative importance of the impact categories. However, not all the impact categories are interrelated so that they can be compared readily. Often the weighting of impact categories can be an analysis based on stakeholders' policy which can be reportable (World Business Council for Sustainable Development (WBCSD) and World Resources Institute (WRI), 2011). The calculation procedure is as follows, while the unit is year/dollars/eco-points or millipoints:

Weighting Index = Weighting factor per category $\times$ Indicator results per group.
Example of elements within life cycle impact assessment:
Impact category: GWP.
Category indicator: kg $CO_2$-eq for GWP.
Characterisation model: ILCD.
Classification: $CO_2$ and $CH_4$ for GWP.
Characterisation: (characterisation factor $\times$ $CO_2$ released) + (characterisation factor $\times$ $CH_4$ released) = GWP.

## Results Interpretation

Results relating to the objectives of the LCA study (ISO 14044: Environmental management − Life cycle assessment − Life cycle interpretation). This phase consists of the identification of critical environmental issues, evaluation of the analysed results, sensitivity analysis, conclusions, and future recommendations. By definition, it is the identification and evaluation of impact assessment results and presents the results in the form of goal/scope/functional unit/system boundary of the analysed system. This is the way of communication among life cycle analysts and community (Ministry of Infrastructure and Environment, 2011). If the interpreted results fail to meet the criteria, then revision of life cycle inventory should be conducted by improving the datasets collected or recalculating the characterisation/normalisation/weighting phases. It can also be supplemented by sensitivity analysis or uncertainty analysis for process improvement. The last step is to check for completeness of an LCA study, which means identification of incomplete datasets/sensitivity check of system or inventory to identify the influential parameters. In summary, the major step includes the evaluation for completeness check, sensitivity check and consistency check (European Commission - Joint Research Centre - Institute for Environment and Sustainability, 2010a).

## Advantages and Limitations of LCA

The advantages of LCA includes:

- LCA systematically analyses the environmental impacts of an entire production system from raw material acquisition to the final disposal.
- LCA allows for hotspot analysis to improve the product or process.
- LCA helps to avoid the problem shifting from one process to another within a product system.
- LCA is a flexible tool which can be combined with other tools like life cycle costing analysis, material flow analysis, environmental impact assessment, and environmental accounting and multicriteria decision analysis.
- LCA is a decision support tool considered as an indicator for sustainability valuation of a company.
- LCA is useful for ecolabelling, which is the marketing of products, while achieving environmental objectives.
- LCA allows comparison among different products or enterprises.

The limitations of LCA are

- The accuracy and time-related framework for LCA vary to a larger extent.
- LCA is data-intensive, which is dependent on geography, data quality, and data availability.
- LCA requires a high level of expertise, while it does not consider technological changes for the future directly.

## Impact Categories

### Acidification Potential

- Impacts on air.
- Impact category that estimates the emissions while increasing the acidity of water and soils.
- The most common form of deposition is acid rain, dry and cloud deposition, excluding the ocean acidification.
- This impact can be varied regionally, while only anthropogenic and natural sources are included.
- The commonly reported impact indicator is kg $SO_2$-eq or mol $H^+$-eq.
- Primary sources are electricity and transportation from fuel combustion and agriculture.
- Main chemical substances are $NO_X$, $SO_X$ and $NH_3$.

- Midpoint indicator − increased acidity in soil and water.
- Endpoint indicator − impacts on organisms, plants and buildings.

## Global Warming Potential

- Impacts on air.
- Impact category that estimates the increase in greenhouse gas concentrations, while increasing global average surface temperature due to greenhouse gas effects. Alternatively called 'Climate Change'.
- An impact indicator is kg $CO_2$-eq. They are generally reported in 100-year time scale.
- Primary sources are electricity and transportation from fuel combustion, agriculture and industrial processes.
- Main chemical substances are $CO_2$, $CH_4$, $N_2O$, $O_3$, $H_2O$ and CFCs.
- Midpoint indicator − increased radiative heating.
- Endpoint indicator − impacts on sea-level change, weather variability, illness due to heat change, wind and ocean.

## Ozone Depletion Potential

- Impacts on air.
- Ozone is the molecule of three oxygen atoms which is colourless and odourless gaseous substance.
- Impact category that estimates the reduction of ozone concentration in the stratosphere, which is good ozone to reduce ultraviolet radiation.
- The primary reason for this impact is replacement or change of chemical composition of CFCs and halons.
- Impact indicator is kg CFC 11-eq.
- Primary sources are manufacturing of polymers or aerosols, fire extinguishers and refrigerant systems.
- Main chemical substances are CFC-11, CFC-12, HCFC-22 and Halon-1301.
- Midpoint indicator − decrease in stratoscopic ozone concentration.
- Endpoint indicator − crops damage, damage to human skins and marine life.

## Smog Creation Potential

- Impacts on air.
- Impact category that assesses the formation of ground-level ozone. Also named as photochemical ozone creation potential. Impacts are relative to sunlight, air quality and population.
- An impact indicator is kg $NO_X$-eq, kg $C_2H_4$-eq or kg $O_3$-eq.

- Primary sources are energy generation, industrial processes and vehicles.
- Main chemical substances are $NO_X$, $O_3$ and $C_2H_4$.
- Midpoint indicator − formation and increase of ground-level ozone.
- Endpoint indicator − Reduced lung function, asthma and irritation in the eye.

## Eutrophication Potential

- Impacts on air, water, and soil.
- Impact category that estimates the biological activity of organisms in excess due to overnutrition. Impacts largely on marine plants due to too many nutrients.
- An impact indicator is kg $NO_3$-eq, kg $PO_4$-eq, kg P-eq or kg N-eq.
- Major sources are wastewater, and fossil fuel combustion.
- Main chemical substances are $NO_X$, $NH_3$, nitrogen, and phosphorus.
- Midpoint indicator − algae growth or marine plants growth excessively.
- Endpoint indicator − biodiversity loss and loss of marine aquatic life.

## Human Toxicity Potential

- Impacts on air, water and soil.
- Impact category that estimates the impact on human health which can be carcinogenic or noncarcinogenic. Caused due to organic chemical and metal compounds.
- An impact indicator is CTU (Comparative Toxicity Unit), kg Benzene-eq or kg Toluene-eq.
- Major sources are mining, manufacturing, energy generation and agriculture.
- Main chemical substances are arsenic, benzene, chromium, dioxins, formaldehyde and zinc.
- Midpoint indicator − human health effects.
- Endpoint indicator − asthma, cancer and heart disease.

## Ecotoxicity Potential

- Impacts on air, water and soil.
- Impact category that estimates the impact on ecosystems which can decrease the production or increase the losses of biodiversity.
- An impact indicator is CTU (Comparative Toxicity Unit) or kg 2,4-dichloro phenoxy-acetic acid eq.
- Major sources are mining, manufacturing, energy generation and agriculture.
- Main chemical substances are organic chemicals, copper, and zinc.

- Midpoint indicator − degradation of ecosystems.
- Endpoint indicator − decreased population and biodiversity.

## Particulate Matter Formation

- Impacts on air, water and soil.
- Impact category that estimates the impact on human health due to the respiration of very small particles.
- The impact indicator is kg $PM_{2.5}$-eq, kg $PM_{10}$-eq or Disability-Adjusted Life Year.
- Major sources are fossil fuel combustion, burning of wood and dust.
- Main chemical substances are $PM_{10}$, $PM_{2.5}$, $SO_X$ and $NO_X$.
- Midpoint indicator − human exposure to particulates.
- Endpoint indicator − asthma, cancer and heart disease.

Chapter 2

# Life Cycle Assessment in Mining Industries

## Introduction

Mining and mineral processing industries contribute a gigantic share in a nation's economy in addition to supplying invaluable resources for modern civilisation. Most importantly, this sector plays a crucial role in sustainable economic development and overall GDP growth. To keep pace with the growing demand of the sustainable world, the World Economic Forum is aiming to make the mining world sustainable by 2050. There are a few goals which should be achieved, and several industrial mining governing bodies are working to settle those goals. These are drivers of change, transitional areas, resource scenario in future and action plans. The drivers of change can be classified into five types: environmental, technological, societal, geopolitical, and geographic for the growing need in the environmental effect's management in the field of climate change, biodiversity and water. World economic forum also outlined the key mining and mineral sectors, which are crucial for setting the sustainable development goal by 2050: aluminium, iron, nickel, copper and zinc mining (Ranängen and Lindman, 2017; World Economic Forum, 2015). In line with the economic demand and growth, the environmental effects and its consequences are also increasing exponentially. Sustainable development meant to an accord between economic activity and environmental concerns. The goal of sustainable development in mining industries is to increase metal production in such a way that it should ascertain the cost and efficiency, without significantly reducing the potential for the future generation. The first step to achieve this goal is to assess the environmental impacts caused by the mining industries for current practices and identify the ways to reduce the environmental effects without compromising the production. According to the International Council on Mining and Metals, there are 10 principals, of which two are directly related to climate change, water and biodiversity — the considerations of the environmental side of sustainable development. According to principle number six,

Life Cycle Assessment for Sustainable Mining. https://doi.org/10.1016/B978-0-323-85451-1.00002-0

continuous improvement should be pursued in environmental issues like water, climate change, and energy use (Gorman and Dzombak, 2018; Mudd, 2009; Tost et al., 2018).

About the environmental impacts of mining and mineral processing, it can be generally classified as waste management, acid mine drainage, sedimentation, metals deposition and biodiversity. The processes produce a considerable amount of waste outputs, which depends on the type of mineral to be mined and type of ore deposit. It is very challenging to dispose of such a large amount of wastes from mining and mineral processing which is detrimental for the aquatic environment and ecosystems. Water pollution from mining wastes causes acid mine drainage, sedimentation and deposition of metals. Erosion through waste rock piles and runoff after rainfall increases the sedimentation (Zhao et al., 2019). Acid mine drainage is another type of serious environmental impacts from mining, which occurs from sulphidic minerals. Contamination of acidic water on surface and groundwater resources threatens the aquatic life and plants. Metals and reagents used for processing the significant amount of minerals also get released into the environments, get mixed, and threaten the life of aquatics and plants. Removing the plants and vegetables from the ore mining area results in deforestation and threatens the structure of the species, which is affecting the biodiversity (Cortez-Lugo et al., 2018; Fugiel et al., 2017). Though the degree of these environmental impacts varies from one mine to another, from one metal to another metal, and correlates with the volume of production, the need for determining and quantifying the effects caused by the sector is inevitable to identify the key processes/materials which are accountable for major environmental emissions and impacts. For example, the key processes are power plants used for electricity generation, diesel used for process heat generation, blasting process or the ore mining (Farjana et al., 2018b,c). Among the various factors, fossil fuel consumption for electricity generation and process heat generation is the greatest contributor to environmental effects and emissions caused by the mining industries. The form of nonrenewable fuels includes coal, natural gas, diesel, heavy fuel oil, and bitumen which are producing heat and electrical energy. After combustion, fossil fuel generates carbon dioxides, nitrous oxides and sulphur dioxides, and produces fly ashes which are polluting the environment. Due to the increasing demand for global development, metal exploration is going into deeper of the surface, which also requires heavy equipment to extract lower-grade ore minerals. These lower-grade metal extractions also demand higher consumption of electricity for more equipment in use. However, increased environmental emission should be diversified in the fields of the environment — global warming, human health, ecosystems and resources (Farjana et al., 2018b,c, 2019a).

In the mineral processing industries, in the last 15 years, 40 significant research outputs are published based on the life cycle assessment (LCA) of specific metal mining processes or a system of processes. The LCA results vary from one standardised LCA method to another, from one simulation software to another and from one mining metal to another. Despite all the key factors, the liable processes crucial for environmental emission for each metal are fixed irrespective of the quantification of results. This statement is validated using the review presented in this chapter. The main aim of this chapter is to quantify and analyse the past research in LCA of minerals and mining processes to fill the knowledge gaps in major mining and mineral production processes and materials accountable for environmental impacts from mining, environmental categories to be affected, the quantity of emissions, and suggest alternatives to reduce the environmental impacts from the mining processes.

## Analysis Methodology

As the first step of this systematic review, the relevant and significant research outputs and the published articles are sorted and classified as part of the material collection process. In total, 40 articles are exclusively selected, which are focussed on the LCA of mining industries. Among which most of the papers were published in *Journal of Cleaner Production, International Journal of Life Cycle Assessment, Science of the Total Environment, Resources Conservation and Recycling, Journal of Environmental Management,* and *Journal of Sustainable Mining.* These papers are classified based on the type of metal mining industry under consideration. Overall 16 mining processes were rigorously studied through LCA. Those are aluminium (3 papers), copper (5 papers), coal (4 papers), gold (3 papers), iron (7 papers), rare-earth element (2 papers), uranium (4 papers), zinc (3 papers), nickel (4 papers), cemented carbide (1 paper), ferroalloy (1 paper), manganese (1 paper), magnesium oxide (1 paper) and titanium oxides (1 paper).

In the next step, these publications are extensively studied to identify the key criteria for conducting a comprehensive literature review. This step can be termed as a comprehensive analysis of the existing literature. In this stage, all the publications are studied irrespective of their evaluation methods, goal and scope of their research. This stage also investigated the common trends and evaluation criteria followed by mining LCA specialists. Furthermore, LCA methods that were considered in those studies are also classified. LCA methods are summarised as their impact assessment methods in practice, system boundary, impact categories in use, general material inputs and outputs, databases and their functional units (Raugei and Winfield, 2019). In the following step, the evaluation categories are selected, and the papers, their methods, and results are sorted based on the evaluation categories. The criteria are selected based on the metals, their general properties, mining processes, system boundary, geographic location, selected environmental impact categories to be assessed, most impactful mining process identification and key emission materials. In the last step, materials and their evaluated criteria were scientifically evaluated, presented and identified the research gaps in these previous research studies. The papers are further studied to find out the research gap on metals which got comparatively lesser attention from LCA practitioners, mining processes which lack attention, key materials used in mining which are harmful to the environment, and suggesting possible solutions to reduce the notable amount of emissions from the mining process or industries. Fig. 1 describes the key steps of the methods utilised to perform this review study.

## Goal and Scope Definition in LCA of Mining

This section describes the basics of LCA, steps, goal and scope, system boundary, software in use and methods used for LCA analysis focussed in the mining industries so far.

LCA is a widely used environmental impact analysis methodology which is used for many years in the scientific community to analyse the effects caused by a product, a process or a system of processes on the environment (Grande et al., 2017). This impact analysis method aims to analyse and classify the environmental emissions based on several categorised and standardised impact assessment categories (Curran, 2012). The major steps in the LCA comprise of goal and scope definition, LCI (life cycle inventory) modelling, LCA analysis, and results interpretation. These steps should be followed in each LCA study conducted based on the ISO (International Organization for Standardization) 14040 (Fogler and Timmons, 1998; Mahmud et al., 2018c; Mahmud et al., 2019). The system boundary followed in an LCA study could

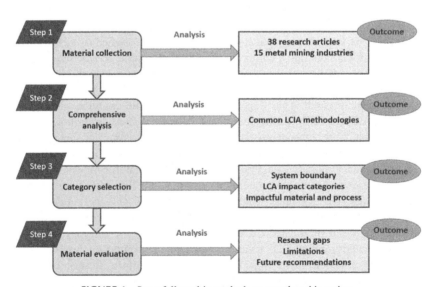

**FIGURE 1** Steps followed in methods to complete this review.

be from cradle-to-gate, gate-to-grave, gate-to-gate, or cradle-to-grave (Stewart and Petrie, 2006). The system boundary followed in mining-based LCA studies in this chapter are cradle-to-gate. Due to the lack of enough data source, it is hard to assume the end-of-life state of the processed metal, which, in turn, restricts the LCA study onto the cradle-to-gate processes (Santero and Hendry, 2016). A complete LCA study in mining comprises of ore mining, concentration and beneficiation, extraction, smelting, transportation, and refining operations. However, the structure of the system boundary is also dependent on the available datasets, as they are proprietary information. The functional unit for LCA in mining could be of 1 kg of selected mineral, or 1 tonne of selected mineral, or a mega tonne of that selected mineral. The goal of the LCA works is to analyse the effects of a product or process based on their system boundary and categorise their emissions (Teh et al., 2017): the materials, energy (in the form of renewables and nonrenewables, resources, organic and inorganic chemicals are identified as the material inputs in an LCA study) (Mahmud et al., 2018a,b,c,d). The output product, waste emissions, by-products and emissions to soil, water and environment are considered as material or process outputs (Althaus and Classen, 2005).

## Life Cycle Inventory Analysis

The LCI datasets were collected from several sources like company reports, published literature, renowned databases like EcoInvent, USGS, and AusLCI. However, very few datasets are focussed on global context. Datasets can be presented as global datasets, regional datasets, or country-specific datasets. For standardising the datasets, the collected sets of data are quantified and aggregated, validated, averaged and finally represented in the form of mine irrespective datasets. The source of these datasets could be originated from mining company reports, published research works, books, media and so on (EPA, 2018). A global dataset could be generated after data collection from several mining companies of different geographic regions, aggregating, averaging, checking and data validation. Most of the metal LCI datasets are regional, which is confusing because technology mix, heat mix and grid mix vary from one company to another, one country to another, and from one region to another. Another major difference is most of the metals could be mined in alternative routes and alternative techniques. It is hard and not available to get the inventory datasets for each of those mining processes.

In most cases, inventory datasets for subprocesses of the full mining process were available, which made it harder to compare among the mining processes and the metals. Also, the metal mining industry datasets are dispersed in different databases which also difficult for researchers to get the LCI datasets in a go. These differences necessitate the fact that in future, it would be highly beneficial to develop global datasets and database containing global metal mining datasets (Althaus et al., 2007; Althaus and Classen, 2005; Hischier et al., 2010; Long et al., 1998; Marguerite et al., 2015; Weidema et al., 2013).

Regarding geographic region, many studies were focussed on different regions, but most commons were based on two countries — Australia and China. Many research on LCA is being carried out in Australian geographic context — aluminium, coal, copper, ferroalloy, gold, iron, nickel, titanium oxides and Uranium. For mining in China, LCA on aluminium, coal, iron and zinc is most commonly assessed. Impact categories were also different based on metal specific industries, as they were originated from different geographic context and used different LCA methods (Mutchek et al., 2016; Stewart and Petrie, 2006; Teh et al., 2017). Allocation technique was required in many of the LCA studies which produce a significant number of by-products. The allocation technique could be of mass-based, economy-based or can be used as a combination of both mass and economy. The characterisation and normalisation factors for allocation techniques vary among the LCA software like SimaPro or Gabi. According to the trend followed by the metal mining industries, it is best practice to avoid coproduct allocation whenever possible. It was assumed that no products were going outside the system other than the main product. However, in many processes, it cannot be avoided. In those cases, it is better to use the allocation technique focussed on both mass and economy, as the mass can vary from one mining company/site to another, production technique could also vary. Also, results between mass and economy could vary in a large amount. Hence, it is best to calculate using both techniques, and then data aggregation and averaging could be done to allocate proper values (Raugei and Ulgiati, 2009; Santero and Hendry, 2016; Weidema and Norris, 2002). The software which was most widely used in mining industries were SimaPro and Gabi. AusLCI database which is integrated with SimaPro software and containing many metal mining LCA datasets focussed particularly for Australia made SimaPro a widely accepted choice. The analysis results generated by both SimaPro and Gabi should produce reasonably similar outcome considering similar analysis techniques and databases are used. However, that is also subject to research and validation. There may be some variation among the datasets generated from different companies; the characterisation and normalisation factors would also be different (Goedkoop et al., 2014; Nunez and Jones, 2016; PRé, 2018; Zhang et al., 2018).

## Life Cycle Impact Assessment Methods

The methods to conduct the LCA are based on ISO 14040 standards. The renowned and widely used LCA methods in mining industries are CML (Center of Environmental Science of the Leiden University of Sweden) method, TRACI (Tool for the Reduction and Assessment of Chemical and other Environmental Impacts) method, ILCD (International Reference Life

Cycle Data System) method, CED (Cumulative Energy Demand) method and IPCC (Intergovernmental Panel on Climate Change) method. These methods vary from one another regarding characterisation, normalisation and weighting factors, impact categories and geographic location. The analysed impact categories also vary among the methods. The most common midpoint indicator-based (problem-oriented) impact categories are HH (human health, cancer and noncancer effects), CC (climate change), eutrophication (terrestrial, marine, aquatic), acidification potential (AP), resources depletion, land and water use, and ecotoxicity. The endpoint indicator (damage-oriented) impact categories are climate change, human health, ecosystems and resources. These methods are also geography dependent like CML methods are widely used for manufacturing processes held in Europe, whereas TRACI method is used for LCA studies based on the United States (Acero et al., 2015; Hischier et al., 2010; JRC European commission, 2011; Wolf et al., 2012).

The most common LCA methods utilised in mining industries were the ReCiPe method, the IPCC method and Australian Life Cycle Indicator based method. ReCiPe method has 18 midpoints indicator-based categories and three end-points indicator-based categories. The ReCiPe is the most updated LCA method with a minimum number of indicator scores. ReCiPe method is classified on hierarchies, individualist and egalitarian method (Fugiel et al., 2017; Marguerite et al., 2015). IPCC method is specialised in greenhouse gas emissions (GHGs). This method accesses the GHG effects based on 20, 50 and 100 years' time span (Hischier et al., 2010). On the other hand, Australian Life Cycle Indicator based method is specialised in materials and processes located in Australian geographic region. According to the best practice guide in LCA, Australian Indicator method does not contain the toxicity emission factors, and the characterisation factors are based on the European context, inherited from the CML method (Lodhia and Hess, 2014; Marguerite et al., 2015). CML method, developed by the University of Leiden in Netherland, is another well-established LCA method, where the characterisation factors are focussed on the average values of global and European geographic context, with no weighting applied on it. So, the regional validity of this method could be considered as global except for the AP and POCF, which are based on European factors. ILCD method is developed by the Joint Research Commission in Europe (JRC) where the impact categories are formed using the base of different renowned methods, IPCC (for climate change in 100 years' time span), USEtox (human toxicity (HT) cancer and noncancer), CML 2002 (resource depletion) and several other methods from literature. This method follows best practice in several impact assessment categories (European Commission − Joint Research Centre − Institute for Environment and Sustainability, 2010; JRC European Commission, 2011). The most discussed impact categories are CC/GHG and PED/TER. AP, MD, waste was also assessed in many literature works, but still, there are huge lacking regarding environmental impact categories. Due to the lack of complete LCI datasets, it was also hard to complete full LCA study for metal industries, which could be a great research focus in future. Also, as environmental impacts could not be regional, there is still necessity to develop a global LCA method, which would apply to processes or system of processes for any region (Acero et al., 2015; Menoufi, 2011). Choosing the appropriate LCA method is a crucial step in conducting an LCA. However, this choice is dependent upon several factors such are geography, characterisation and normalisation factors for the environment and industrial sector choice. In mining industries, the most commonly practised method is the ReCiPe method and ILCD method (discussed in Table 2).

**TABLE 2** Summary of LCA techniques used in previous studies.

| Metal | Method | Impact categories |
| --- | --- | --- |
| Aluminium (Farjana et al., 2019a; Nunez and Jones, 2016; Paraskevas et al., 2016; Tan and Khoo, 2005) | EDIP, UMIP, ReCiPe, ILCD, TRACI | GWP, AP, HT, resources, waste, FFD, EU, ODP, POCP, WF |
| Cemented carbide (Furberg et al., 2019) | | TE, OD, FEW (freshwater ecotoxicity), CC, POF, WD |
| Coal (Adiansyah et al., 2017; Burchart-Korol et al., 2016; Guimarães da Silva et al., 2018; Zhang et al., 2018) | IPCC, ReCiPe, Australian indicator | AP, GWP, resource, waste, dust |
| Copper (Ekman Nilsson et al., 2017; Haque and Norgate, 2014; Memary et al., 2012; Norgate, 2001; Northey et al., 2013) | CSIRO software | GWP, AP, TED |
| Ferroalloy (Bartzas and Komnitsas, 2015; Haque and Norgate, 2013) | | GWP, AP, PED |
| Gold (Chen et al., 2018; Haque and Norgate, 2014; Norgate and Haque, 2012) | ReCiPe, Australian indicator | Energy, GWP, water use, waste, TA, HT, PMF, ME, FD |
| Iron (Ferreira and Leite, 2015; Gan and Griffin, 2018; Haque and Norgate, 2015; Norgate and Haque, 2010) | IPCC, Australian indicator | GHG, human health, ecosystem, resources, GER |
| Manganese (Westfall et al., 2016) | CML 2001 | GWP, AP, POCP, water use, waste |

**TABLE 2** Summary of LCA techniques used in previous studies.—cont'd

| Metal | Method | Impact categories |
|---|---|---|
| Magnesium oxide (Ruan and Unluer, 2016) | EcoIndicator 99 | Human health, ecosystem, resources |
| Nickel (Khoo et al., 2017; Mistry et al., 2016) | IMPACT 2002+, ReCiPe, Australian indicator, CML 2001 | GWP, PED |
| Rare-earth element (Lima et al., 2018; Zaimes et al., 2015) | IMPACT 2002+, ReCiPe, CED | GER, GWP, ODP, carcinogens |
| Silver (Farjana et al., 2019b,d) | ILCD, CED | GWP, HT |
| Steel (Burchart-Korol, 2013; Norgate et al., 2007; Renzulli et al., 2016) | IPCC, ReCiPe, CED, ILCD | GWP, AP, TED, dust, ODP, PMF |
| Titanium oxides (Farjana et al., 2018c) | ILCD, CED | GWP, AP, FAE, MAE, EU, MRD, LU, water, TED |
| Uranium (Farjana et al., 2018e; Haque and Norgate, 2014; Norgate et al., 2014; Parker et al., 2016) | ILCD, Australian indicator, CED | GWP, AP, FAE, MAE, EU, MRD, LU, water, TED |
| Zinc (Qi et al., 2017; Van Genderen et al., 2016) | ReCiPe, CML 2001 | GWP, AP, EU, POCP, OD, PED |

The identification of the best practice methods under different impact categories are discussed below in Table 3. Table 3 also elaborates the quantification techniques for each of the major environmental impact assessment categories.

## Results Analysis based on Metal Mining Industries

This section describes the LCA conducted on the key mining and mineral processing industries, their analysis methods, key findings, results, limitations, and future recommendations. The major industries involve aluminium, coal, copper, ferroalloy, gold, iron, nickel, rare-earth elements, stainless steel, uranium and zinc. The last subsection identifies five metals with no or very few research output which is cemented carbide, manganese, magnesium oxide and titanium oxides.

**TABLE 3** Quantification and best practice in LCA methods (Marguerite et al., 2015).

| Impact categories | Elements used for quantification | Best practice methods |
|---|---|---|
| Climate change | Quantification based on human activities on climate based on greenhouse gas emissions. This is most commonly accounted for carbon dioxide, methane and nitrous oxide emissions. | IPCC method to calculate GHG based on 100 years' emissions. |
| Resources depletion | Quantifies the depletion of natural resources from the earth, based on the concentration of reserve and deaccumulation rate/quantity of fuels/ratio of annual production to available reserve/damage to resource-based on the increased cost of extraction. | CML method and the ILCD method based on the concentration of reserve and deaccumulation rate. |
| Eutrophication | Quantification based on the macronutrients released on the environment — air, water, soil. It can be aquatic and terrestrial. Aquatic eutrophication is quantified based on accelerated algae growth, reduced sunlight infiltration and oxygen depletion. Terrestrial eutrophication is quantified based on increased susceptibility of plants to diseases. | CML and IMPACT 2002+ which quantifies based on stoichiometric nutrification potential applicable to both categories. |
| Acidification | Quantification based on the acidifying impacts based on when acid precursor compounds are released onto the environment and deposited on land (terrestrial) or water (aquatic). Quantification is mainly based on nitrous oxides, sulphur oxides, sulphuric acid and ammonia. | CML and ILCD method based on the critical load exceedance method of hazard index method. |
| Human toxicity and ecotoxicity | Quantifies the impact of toxicity substance released on land, water and environment. Quantification is based on using pesticides, heavy metals, hormones and organic chemicals. | USEtox, ILCD, ReCiPe and IMPACT 2002+. |
| Photochemical ozone formation | Quantifies the impacts based on impacts from the increase in ozone formation in troposphere. The main criteria are the emission of nitrous oxides, carbon monoxide and those which impacts on ozone formation. | CML method, based on the simplified description of atmospheric transport. |

**TABLE 3** Quantification and best practice in LCA methods (Marguerite et al., 2015).—cont'd

| Impact categories | Elements used for quantification | Best practice methods |
|---|---|---|
| Particulate matter formation | Quantification is based on the emissions on air which are harmful to human health. In different LCIA methods, these are characterised by different impact categories. | TRACI as TRACI method distinguishes between different types of emissions. |
| Land use | Quantification is based on the amount of land use in LCA and its effects on biodiversity. | Currently, no best practice methods for land use as no single method quantifies all levels of biodiversity. |
| Ozone depletion | Quantifies the impact based on the reduction in the concentration of ozone. | Ozone depletion factors published by the World Meteorological Organizations. |
| Ionising radiation | Quantifies the impact of radioactive species (radionuclides) on air and water. | ILCD or ReCiPe method based on the quantification of radioactive impact on human health. |

## LCA in Aluminium Mining

Aluminium is a lightweight and corrosion-resistant metal, which are used in automobile industries, aerospace industries, beverage, making alloys and electronics industries. The top countries which are mining and producing aluminium are China, Russia, Canada, India and Australia. Aluminium production consists of four major steps: bauxite ore mining, alumina production from Bayer process, alumina smelting from Hall Heroult process and alumina refining. Table 4 describes the major findings and recommendations from the researchers conducted related to LCA of this metal.

There are four major studies conducted in the LCA of aluminium production processes. Tan et al. analysed LCA of the aluminium supply chain which consists of the refinery, smelter and a casting plant. The geographical coverage was Australia, and it was a cradle-to-gate LCA study. The analysis was done using SimaPro with EDIP (Environmental Design of Industrial Products). The impact categories which were assessed are global warming potential (GWP), AP, HT, resources and built waste. Tan considered four scenarios, including the baseline mode of operation, and other scenarios were constructed after modifying the scrap metal and red mud. Their results

**TABLE 4** Summary of the findings and recommendations for the LCA of aluminium mining.

| Study reference | Key findings | Recommendation |
| --- | --- | --- |
| LCA focussed on Australia (Tan and Khoo, 2005) | Electricity use in the smelting and refining process is mainly responsible for GWP. | Reducing the scrap metal and red mud could significantly reduce the impacts. |
| LCA comparing 29 countries (Paraskevas et al., 2016) | Identified that countries with high coal and oil-rich mix have higher emission results and countries with similar electricity mix show similar GHG results. | To reduce the environmental impacts from aluminium production, it should start by focussing on China due to their energy mix. All the aluminium-producing countries should focus on using cleaner resources. |
| LCA on global aluminium production (Nunez and Jones, 2016) | Reveals the fact that the highest contribution of GHG is from alumina refining and electrolysis process. Direct emissions from the electrolysis process also contribute largely to GHG. | Due to increased production of aluminium in China, more LCI datasets are required for modelling to enhance the reliability of the impact assessment about aluminium production. |
| LCA on global aluminium production (Farjana et al., 2019a) | Electricity and process heat consumption in alumina smelting has the highest impact. | Alternative energy generation source and improvement in energy efficiency would be beneficial for sustainability. |

revealed that electricity use in the smelting and refining process is mainly responsible for GWP and HT. AP is due to sulphur dioxide ($SO_2$) generation from the power plant and transportation system. Their study revealed that reducing the scrap metal and red mud could significantly reduce the impacts of aluminium production (Tan and Khoo, 2005).

Paraskevas et al. analysed aluminium production after comparing the electricity mix of 29 countries. Their datasets are originated from EcoInvent and USGS databases, whereas analysed using SimaPro software and ReCiPe method. Electricity generated from hydropower, coal, gas, nuclear, biofuel, geothermal and solar is considered there. They assumed that electricity is 100% grid dependent and not traded among the countries. According to Paraskevas, amount of GHGs from aluminium production is 0.45 Gt $CO_2$ eq. Aluminium production requires 66 MJ energy per kg. More than 80% is for electricity generation in Hall Heroult process. They identified that countries with high coal and oil-rich mix have higher emission results rather than countries using cleaner technologies. They have also stated that countries with similar electricity mix show similar GHG results (Paraskevas et al., 2016). Another study conducted by Nunez et al. where the datasets are collected by the International Aluminium Institute (IAI) has studied primary aluminium production process and used Gabi software. The geographic region under consideration was global and global, minus China. Nunez considered six midpoints indicator-based categories: GWP, AP, FFD (fossil fuel depletion), EU (eutrophication), ODP (ozone depletion potential), POCP (photo-oxidant creation potential). Nunez also analyzed the WSP (water scarcity potential). Their identification also reveals the fact that the highest contribution of GHG is from alumina refining and electrolysis process. Direct emissions from the electrolysis process also contribute largely to GHG (Nunez and Jones, 2016). Farjana et al. studied the cradle-to-gate complete LCA of four major steps from the bauxite mining to alumina refining processes. They have also conducted a sensitivity analysis among the quantity of fossil fuel used based on process heat generation. As the study reports, electricity consumption during alumina smelting has the highest impact on the environment. Process heat generation also has a considerable impact on the environment. The climate change effect from bauxite mining is 0.079 kg $CO_2$ eq., alumina production is 1.23 kg $CO_2$ eq., alumina smelting is 10.91 kg $CO_2$ eq. and alumina refining is 0.27 kg $CO_2$ eq (Farjana et al., 2019a).

## LCA in Coal Mining

Coal is a widely used fossil fuel used for electricity generation or process heat generation. Coal is extracted on the earth crust based on opencast and underground mining. The major miners of coal are Ukraine, Kazakhstan, South Africa, China, the United States and Australia. Coal can be produced from open-pit mining or underground mining. There are a few steps involved in coal production, while major processes are mining, transportation and comminution. In the mining step, surface soil is removed for drilling and detonation. Then the inert materials are removed, and coal layers are drilled. Then the coal layers underwent detonation and mined areas are recovered. Then the run of mines undergoes the loading and transportation. The run of mines is then stocked, crushed and shifted. Table 5 describes the major findings and recommendations from the research conducted related to LCA of this metal. Korol et al. developed and utilised a computational LCA for coal mining in Poland. They assessed the GHG using the IPCC method (20, 100 and 500 years). Analysis of damage categories was done using the ReCiPe method. Their study identified that the largest contributor to GHG is a fossil fuel and methane emissions for electricity use, processing of waste, heat and steel supports. Among the three ranges of years, the highest range of emissions was from 20 years' span, that is, 85 kg $CO_2$ eq per kg. In this time frame, the highest contributor for GHG is methane emissions which could be generated

**TABLE 5** Summary of the findings and recommendations for the LCA of coal mining.

| Study reference | Key findings | Recommendation |
| --- | --- | --- |
| LCA for coal mining in Poland (Burchart-Korol et al., 2016) | The largest contributor to GHG is a fossil fuel and methane emissions for electricity use, processing of waste, heat. | The impacts can be reduced using alternative fuels in steel production. Pollution prevention methods should be applied to reduce emissions. |
| LCA in respect of Australia (Adiansyah et al., 2017) | Electrical energy is the greatest contributor to environmental emissions from coal mining. | The thickened tailing in the coal mining tailing management could reduce environmental burdens. |
| LCA focussed on coal production in China (Zhang et al., 2018) | The most impactful category was dust, followed by GWP and acidification. | Increase in tire performance, transportation efficiency, low-carbon power development and advancement in mining technology will be beneficial to reduce the impacts. |
| LCA based on Brazil (Guimarães da Silva et al., 2018) | Diesel oil is the significant parameter which emits carbon dioxide, methane and ammonia. | This work will contribute to the detailed LCA analysis of coal mining in Brazil to implement the measures related to diesel oil. |

from electricity, heat and steel support associated with the ventilation process (Burchart-Korol et al., 2016). In another study, Adiansyah et al. compare coal mining tailings management strategies using LCA. They have used hybrid LCA method comprised of Australian Indicator and ReCiPe method. The strategies involved belt press, tailings paste, thickened tailings. They found that electrical energy is the greatest contributor to environmental emissions from coal mining. The analysis is done using three case scenarios — tailings with a low amount of water, tailing paste with much solids and the base case with a high amount of water (Adiansyah et al., 2017). Zhang et al. studied the opencast coal mining operation in China. The functional unit was 100 tonne of coal. The system boundary includes stripping, mining, transportation, processing and reclamation — the impact categories which were considered: resources, acidification, GWP, waste and dust. According to Zhang, the most impactful category was dust, followed by GWP and acidification. For 1 tonne of coal production, the GWP was 7331.7 kg $CO_2$ eq (Zhang et al., 2018). Silva et al. analysed the surface coal mine located in Brazil. Their system boundary includes mining, transportation system, comminution, recovery, production and final transportation. Their study found that diesel oil is the significant parameter which emits carbon dioxide, methane and ammonia. They have also conducted a sensitivity analysis based on fugitive emission factors for diesel oil, electricity and transport (Guimarães da Silva et al., 2018).

## LCA in Copper Mining

Copper is a valuable metal used for its conductivity properties and corrosion-resistant properties. They are widely used for electrical wiring, construction, heat exchangers and electronics. The producers of copper are the United States, Chile and Australia. Copper is extracted using ore mining, beneficiation. If it goes through pyrometallurgical extraction, then it passes through smelting, converting, fire refining and electrorefining. If the ore goes through hydrometallurgical extraction, then it has leaching, solvent extraction and electrowinning. Table 6 describes the major findings and recommendations from the research conducted related to LCA of this metal. Norgate conducted LCA based on pyrometallurgical and hydrometallurgical copper production. They assessed and compared the impacts from process parameters, ore grade, energy consumed for electricity generation and electricity generation capacity. The assessed impact categories were GWP and AP. They showed that solvent extraction and electrowinning processes of hydrometallurgical copper production has a higher impact over pyrometallurgical copper production (Norgate, 2001). Memary et al. conducted a time series analysis of copper production. They examined the environmental effects of copper mining and smelting from 1940 to 2008 through a novel approach. They had conducted a cradle-to-gate LCA study based on datasets from five Australian copper mine. They showed that carbon footprint from all mines ranged from 800 to 1922 kt $CO_2$ eq. They showed that copper mining and milling were the most crucial part to reduce GWP (Memary et al., 2012). Northey et al. conducted LCA of copper mining and stated that average energy intensity ranges from 10 to 70 GJ/t Cu. GHG ranges from 1 to 9 tonne $CO_2$ eq/t Cu. This large variation was due to the form of copper produced, ore grade, sources of fuel and electrical energy (Northey et al., 2013). Haque et al. conducted a comparative LCA study among in situ leaching based production of copper, gold and uranium. Their study was conducted based on Australian Impact method using SimaPro software. They showed that for copper mining, solvent extraction and electrowinning played a significant role in GHG. For good field-related activities, electricity was the main source of GHG and also the sulphuric acid used for leaching (Haque et al., 2014). Nilsson compared the carbon footprint of copper mining based on LCA, while the sources are both primary and secondary. The value of carbon footprint ranges from 1.1 to 8.5 kg $CO_2$ eq/ kg Cu. The literature shows that it could be $2.1E^8$ kg $CO_2$ eq/kg Cu for primary copper. From the secondary source, the carbon footprint ranges from 0.1 to

**TABLE 6** Summary of the findings and recommendations for the LCA of copper mining.

| Study reference | Key findings | Recommendation |
|---|---|---|
| LCA on global copper production (Norgate, 2001) | Solvent extraction and electrowinning processes of hydrometallurgical copper production have a higher impact. | They conducted a sensitivity analysis based on by varying 30% of process power consumption and transport mode and distance, which indicates some improvement in the environmental emissions. |
| LCA on Australian copper mining (Memary et al., 2012) | Copper mining and milling were the most crucial part to reduce GWP. | The mining production impacts related to regional energy consideration should be carefully considered with geography. |
| LCA on copper mining (Northey et al., 2013) | The large variation in GHG was due to the form of copper produced, ore grade, sources of fuel, and electrical energy. | Recommendations made for the companies for improving mining datasets by indicating detailed energy consumption scenario for each stage of the metal production. |
| LCA of copper mining in Australia (Haque and Norgate, 2014) | For good field-related activities, electricity was the main source of GHG, and the sulphuric acid used for leaching. | Electricity generated from renewable energy resources such as solar energy will be beneficial for reducing impacts from in situ leaching. |
| LCA of global copper mine (Ekman Nilsson et al., 2017) | The reason for this wide range of variation in GHG was the material quality, the metallurgical process and transportation distance. | In case of the advancement of the technology for efficient recovery of metals, the impact will reduce. |
| LCA of gold−silver−lead−zinc−copper beneficiation (Farjana et al., 2019b) | Gold−silver beneficiation has the highest impact over the other metals. | Modification in the electricity grid mix to enhance the energy efficiency will be helpful to reduce environmental burdens. |

1.9 kg $CO_2$ eq/kg Cu. The reason for this wide range of variation was the material quality, the metallurgical process and transportation distance (Ekman Nilsson et al., 2017). Farjana et al. analysed the gold−silver−lead−zinc−copper beneficiation process, which summarises that gold−silver beneficiation has the greatest impact over the other metal in these joint production which is due to the electricity usage through fossil fuel consumption (Farjana et al., 2019b).

## LCA in Ferroalloy Mining

Ferroalloys are iron-bearing alloys which also have a higher proportion of metals. They can make alloys including chromium, manganese, molybdenum and silicon. It is mostly found and produced in China, South Africa, Ukraine and Kazakhstan. Table 7 describes the major findings and recommendations from the research conducted related to LCA of this metal. Haque et al. studied the LCA of ferroalloys. These alloys were iron-bearing material alloyed in conjunction with manganese, chromium, silicon and molybdenum. They studied the environmental impacts in respect of Australia, using the Tasmanian electricity GHG emissions factors. According to their results, GHG was 1.8 tonne $CO_2$ eq/t FeMn, 2.8 tonne $CO_2$ eq/t SiMn, 3.4 tonne $CO_2$ eq/t FeSi, 13.9 tonne $CO_2$ eq/t FeNi and 3 tonne $CO_2$ eq/t FeCr. This large variation in GHG was due to their electricity use, fossil fuel consumption and the grade of ferroalloys produced (Haque and Norgate, 2013). Bartzas et al. conducted an LCA study based on ferronickel production with datasets collected from mines in Greece. They assessed GWP, AP and PED based on three different case scenarios. The scenarios were dependent on energy sources and mode of operation in smelting and refining. They showed that environmental impacts

**TABLE 7** Summary of the findings and recommendations for the LCA of ferroalloy mining.

| Study reference | Key findings | Recommendation |
| --- | --- | --- |
| LCA of ferroalloy in respect of Australia (Haque and Norgate, 2013) | The large variation in GHG was due to their electricity use, fossil fuel consumption, the grade of ferroalloys produced. | Renewable carbon like biochar can be blended with fossil fuels like coke or coal to replace these fossils. |
| LCA of ferroalloy in respect of Greece (Bartzas and Komnitsas, 2015) | The environmental impacts could significantly be reduced with renewable energy use instead of fossil fuel. The smelting and refining stage consumes the highest amount of energy, while ore mining and beneficiation contributed the least. | Green energy utilisation would be the best possible solution to get better environmental results in each aspect. |

could significantly be reduced with renewable energy use instead of fossil fuel. In the normal scenario, the GHG was 12.6 tonne $CO_2$ eq/t FeNi, while in renewable energy scenario it could be 8.24 tonne $CO_2$ eq/t FeNi. They also found that the smelting and refining stage consumes the highest amount of energy, while ore mining and beneficiation contributed the least (Bartzas and Komnitsas, 2015).

## LCA in Gold Mining

Gold is a precious metal which is a popular metal of precious jewellery, currency or in electronic applications. Australia has the highest amount of gold mine deposit, similar to South Africa and Russia. From the gold mine, the gold ores are gone through comminution stage comprises of crushing and grinding. If it is refractory ore, then consequentially undergoes flotation, roasting, pressure oxidation, bio-oxidation, regrinding, smelting and refining. If the gold ore is nonrefractory or free-milling ore, then it undergoes an extraction, cyanidation (tank or heap), recovery, stripping, electrowinning, smelting and refining. Table 8 describes the major findings and recommendations from the research conducted related to LCA of this metal. Norgate et al. analysed gold production processes both from refractory and nonrefractory ores. The impact categories under consideration were energy, GHG, water and waste. They found that refractory ore has the highest impact on mining and comminution stage due to electricity consumption. 61% of the total GHG is from refractory ores (Norgate and Haque, 2012). Chen et al. analysed the life cycle impact of gold mining processes in China where the impact categories were CC, TAP, HT, PMF (Particulate Matter Formation) and FFD (fossil fuel depletion). The analysis was done by the midpoint- and endpoint-based ReCiPe method. They found that the largest impact is on the metal depletion category from mining due to energy consumption and emissions. For climate change, electricity and diesel were the largest contributors. Climate change impact was 5.55E4 kg $CO_2$ eq (Chen et al., 2018). Haque et al. conducted a comparative LCA study among in situ leaching-based production of copper, gold and uranium. Their study was conducted based on Australian Impact method using SimaPro software. They showed that for gold mining, activities relating well field impacts mostly about 39% of total GHG, where 56% was from gold extraction and metal production, and 5% were due to chemical use (Haque and Norgate, 2014). Farjana et al. analysed the gold—silver—lead—zinc—copper beneficiation process, which summarises that gold—silver beneficiation has the greatest impact over the other metal in these joint production which is due to the electricity usage through fossil fuel consumption (Farjana et al., 2019b). Farjana et al. also analysed the comparative life cycle environmental impacts of gold—silver refining operations where the ore is extracted from

**TABLE 8** Summary of the findings and recommendations for the LCA of gold mining.

| Study reference | Key findings | Recommendation |
|---|---|---|
| LCA on gold mining in Australia (Norgate and Haque, 2012) | The refractory ore of gold has the highest impact on mining and comminution stage due to electricity consumption. | Gold mining technological improvements would be beneficial for high-pressure gas recovery, leaching, flotation technologies, gravitational technologies, etc., to reduce the environmental impacts. |
| LCA of gold mining in China (Chen et al., 2018) | The largest impact is on the metal depletion category from mining due to energy consumption and emissions. For climate change, electricity and diesel were the largest contributors. | Environmental policy suggestions like increasing the resources efficiency, adjusting energy structure, gold recycling and ecological compensation approach would be beneficial for reducing impacts. |
| LCA of gold mining in Australia (Haque and Norgate, 2014) | For gold mining, activities relating well field impacts mostly about 39% of total GHG, where 56% was from gold extraction and metal production, and 5% were due to chemical use. | Electricity generated from renewable energy resources such as solar energy will be beneficial for reducing impacts from in situ leaching. |
| LCA of gold−silver −lead−zinc−copper beneficiation (Farjana et al., 2019b) | Gold−silver beneficiation has the highest impact over the other metals. | Modification in the electricity grid mix to enhance the energy efficiency will be helpful to reduce environmental burdens. |
| LCA of gold−silver refining operations (Farjana et al., 2019e) | Gold refining from the gold−silver couple production has a notable environmental impact over the combined production. | Altering the material alloying properties for stainless steel would be beneficial to reduce environmental impacts. |

coproduction of gold−silver and the combined production of gold−silver−lead−zinc−copper. Results from that study showed that gold refining from the gold−silver production has a noteworthy environmental impact over the combined production of five metals (Farjana et al., 2019e).

## LCA in Iron Mining

Iron is one of the most abundant elements found in the earth's crust and is the most widely used metal in modern civilisation. The major miners and producers of iron are China, Australia, Brazil, Russia and India. Iron can be mined from open-pit mining or underground mining. Iron ore is extraction through drilling, blasting, excavating, loading, haulage, ore dumping, crushing and screening. After screening lumped ores are sent through stacking, reclaiming, loading and transport. If the iron ores are fine ore, then after screening it goes through stacking, reclaiming, loading, transport, pelleting and sintering. Table 9 describes the major findings and recommendations from the research conducted related to LCA of this metal.

Ferreira et al. analysed the iron production for an open-pit mine located in Brazil. Dataset was collected from company reports. The functional unit considered was 1 tonne of iron ore concentration produced. The analysis was done using SimaPro software and IPCC method. They found that the use of grinding media was the largest contributor to environmental emissions with the largest impact on human health and quality of ecosystems. Emission of inhalable inorganic materials is affecting human health. GHGs are from electricity consumption which was 13.32 kg $CO_2$ eq for 1 tonne of iron produced (Ferreira and Leite, 2015). Haque et al. analysed LCA of open-pit iron ore mining and processing which comprises of drilling, blasting, loading, haulage, crushing, screening, separation, stacking and stockpiling.

**TABLE 9** Summary of the findings and recommendations for LCA of iron mining.

| Study reference | Key findings | Recommendation |
| --- | --- | --- |
| LCA of iron mining in Brazil (Ferreira and Leite, 2015) | The use of grinding media was the largest contributor to environmental emissions with the largest impact on human health and quality of ecosystems. | Replacement of fossil fuels with biodiesel would be beneficial for reducing climate change effects. |
| LCA of iron mining in Australia (Haque and Norgate, 2015) | The loading and haulage made the highest contribution to GHG. | Emerging technologies should be adopted to reduce the environmental burdens from loading and hauling of iron ore. |
| LCA of iron mining in China (Gan and Griffin, 2018) | The highest GHG contribution is from agglomeration, loading, haulage and ore processing. | To reduce the environmental impacts of iron mining in China, high-grade ore import from abroad. |

The functional unit was 1 tonne. The impact categories were gross energy requirements and GHG emissions. The analysis was done by Australian Indicator method. The analysis results showed that loading and haulage made the highest contribution to GHG (11.9 $CO_2$ eq/t) (Haque and Norgate, 2015; Norgate and Haque, 2010). Gan et al. analysed LCA of mining and processing of iron ore mine in China. The iron mine considered consists of open-pit mine and underground mine. The functional unit was 1 tonne of iron ore. The analysis was done using the IPCC method. They found that the highest GHG contribution is from agglomeration, loading, haulage and ore processing, where the mean value is 270 kg $CO_2$ eq tonne. The second-largest contributor is vegetation (Gan and Griffin, 2018).

## LCA in Nickel Mining

Nickel is an important element for the military, marine, transport, aerospace and architectural applications. In the nickel mining, the ore can be extracted from sulphidic ore or oxidic ore. If the ore is sulphidic, then it goes through beneficiation, pyrometallurgical extraction and refining. If the ore is oxidic, then it goes through ore preparation, hydrometallurgical extraction and refining. Table 10 describes the major findings and recommendations from the research conducted related to LCA of this metal. Khoo et al. studied three cradle-to-gate-based nickel laterite technologies. The geographic regions considered were Western Australia and Indonesia. The analysis was conducted using the IMPACT 2002 þ method, ReCiPe method and Australian Indicator method. They were high-pressure acid leaching (HPAL), ferronickel and direct

**TABLE 10** Summary of the findings and recommendations for LCA of nickel mining.

| Study reference | Key findings | Recommendation |
| --- | --- | --- |
| LCA of nickel mining for Australia and Indonesia (Khoo et al., 2017) | The highest environmental impact was from ferronickel, while the most sustainable was HPAL due to fuel consumption in nickel reduction and smelting. | Nickel production using the DNi method would be beneficial over the HPAL method to enhance sustainability. |
| LCA of global nickel production (Mistry et al., 2016) | The extraction and refining steps were the major contributor for PED and GWP. | The nickel institute will update its life cycle inventory datasets based on technological advancements, energy efficiency and raw material inputs. |

nickel (DNi) process. Their study revealed that the highest environmental impact was from ferronickel, while the most sustainable was HPAL. It was also found that results obtained from Australian Indicator method were slightly higher than for other methods. GWP was mostly contributed by fuel consumption in nickel reduction and smelting. Another major factor for GHG was coal consumption to produce steam (Khoo et al., 2017). Mistry et al. conducted the LCA of nickel and ferronickel production. Their datasets were originated from nine companies, which comprised of 52% global nickel production and 40% global ferronickel production. The analysis was done using Gabi software and CML 2001 method. They showed that extraction and refining steps were the major contributors for PED and GWP (Mistry et al., 2016).

## LCA in Rare-Earth Elements Mining

These group of metals contains 15 types of elements, which can exist in the form of oxide minerals. These rare-earth elements are used for making optical and electronics products. Different industrial sectors are using these rare-earth elements metallurgy, oil and agriculture. They are mined from China, Brazil and the United States. Rare-earth elements can be mined using open-pit mining or underground mining. After mining, it goes through beneficiation stage, which involves crushing and grinding, magnetic separation, multistage flotation, filtering, washing and conditioning. At the refining stage, the concentrated rare-earth elements go through leaching, roasting, coproducts separation, ion exchange, calcination, solvent extraction, mixed concentration, mischmetal refining, water leaching, multistage acid leaching, precipitation and acid digestion. Table 11 describes the major findings and recommendations from the research conducted related to LCA of this metal. Weng et al. conducted LCA of rare-earth elements, based on cradle-to-gate 26 mining projects. Their focus was on GER (Gross Energy Requirements) and GWP. The analysis was done using SimaPro software, ReCiPe method, and CED method. The allocation method was an economy-based allocation of SimaPro. They found that the greatest environmental impacts were from ionic clay, monazite, bastnaesite, where the impact is higher than for most common metals like bauxite, copper and steel (Weng et al., 2016). Lima et al. analysed LCA of 4 kg of rare-earth element with 2 kg of coproducts from a Brazilian ore. The analysis was done using the IMPACT 2002 method, where the datasets were from CETEM. Their study showed that the largest consumption of hydrochloric acid and ammonium oxides produces radioactive emissions which impact on ozone depletion potential, and carcinogens thus impact on human health (Lima et al., 2018).

**TABLE 11** Summary of the findings and recommendations for LCA of rare-earth elements mining.

| Study reference | Key findings | Recommendation |
| --- | --- | --- |
| LCA of rare-earth elements (Weng et al., 2016) | The greatest environmental impacts were from ionic clay, monazite, bastnaesite where the impact is higher than for most common metals like bauxite, copper and steel. | Rare-earth elements refining technologies should adopt cleaner ways based on mineralogy and geological conditions. |
| LCA of rare-earth elements in Brazil (Lima et al., 2018) | The largest consumption of hydrochloric acid and ammonium oxides produces radioactive emissions which impact on ozone depletion potential, and carcinogens thus impact on human health. | More investment is required in the life cycle assessment sectors to better justify the environmental impact analysis. |

## LCA in Stainless Steel Mining

The steel industry is like oil and gas. Steel is widely used in construction works, transportation, packaging and energy sector. China is one of the leading producers of steel. Steel is produced after processing in coke ovens, a sintering plant, blast furnace and converter. Table 12 describes the major findings and recommendations from the research conducted related to LCA of this metal. Norgate et al. conducted the LCA study of steel production where the functional unit was 1 kg of refined stainless steel. They assessed GWP, AP and TED. The effects of different sources of electricity were also assessed. They used CSIRO software LCA-PRO for analysis. Their study revealed that when ferronickel was the nickel source, total energy consumption was higher than the case where nickel metal was used as a source. They also stated that nickel has the highest impact, then ferronickel, ferrochrome and iron. While analysing the source of electricity, natural gas use could reduce the GWP without changing the total energy consumption, while hydroelectricity could reduce both (Norgate et al., 2007). Korol et al. assessed the cradle-to-grave LCA of steel production and arc furnace routes of Poland. They used SimaPro software and EcoInvent database for their analysis. The analysis method was IPCC, ReCiPe and CED. They showed that the production of pig iron had the highest impact on GHG and fuel consumption due to electricity. They also showed that the iron ore sintering process was the highest contributor to dust and on GHG emissions. Direct GHG emissions were related to the combustion sources, while the indirect emissions were from fossils. The GHG value was 913 kg

**TABLE 12** Summary of the findings and recommendations for LCA of steel mining.

| Study reference | Key findings | Recommendation |
| --- | --- | --- |
| LCA of steel (Norgate et al., 2007) | Stated that when ferronickel was the nickel source, total energy consumption was higher than the case where nickel metal was used as a source. They also stated that nickel has the highest impact, then ferronickel, ferrochrome and iron. | Alternative metal production technologies should be developed to reduce environmental burdens. |
| LCA of steel production in Poland (Burchart-Korol, 2013) | The production of pig iron had the highest impact on GHG and fuel consumption due to electricity. They also showed that the iron ore sintering process was the highest contributor to dust and on GHG emissions. | In future, LCA of alternative steel production technologies should be conducted, including thermodynamic analysis and exergy analysis. |
| LCA of steel mill in Italy (Renzulli et al., 2016) | The most impactful processes were blast furnace and coke oven operations. More than 40% of the climate change, ozone depletion and particulate matter were due to raw material transportation in the blast furnace and coke oven. | Energy and fuel exchange between the power plants and factory can be modelled and analysed. LCA analysis can also be extended based on steel product rolling operations. |
| LCA of steel production in China (Ma et al., 2018) | The water footprint caused by steel production and found that grey water footprint was higher than blue water footprint, while direct emissions played a key role in grey water footprint. | Further studies should be carried out based on a dynamic database and spatial disparity in water footprint evaluation in different aspects. |

$CO_2$ eq/FU for natural electric arc furnace and 744 $CO_2$ eq/FU for crude steel (Burchart-Korol, 2013). Renzulli et al. studied the cradle-to-gate LCA of an integrated steel mill in Italy. The system boundary was raw material extraction, the sintering operations, the coke production and iron—steel production. The analysis method was ILCD, while both mass and economy-based allocation were considered. Their results are similar like for Korol et al. that the most impactful processes were blast furnace and coke oven operations. More than 40% of the climate change, ozone depletion and particulate matter were due to raw material transportation in the blast furnace and coke oven (Renzulli et al., 2016). Ma et al. analysed crude steel production in China. Their functional unit was 1 tonne of steel billet. The analysis method was IMPACT 2002+ method,

IPCC and ReCiPe. The system boundary includes mining, magnesium oxide production, transportation and electricity generation. They analysed the water footprint caused by steel production and found that grey water footprint was higher than blue water footprint, while direct emissions played the key role for grey water footprint. Metal depletion was higher due to iron ore consumption (Ma et al., 2018).

## LCA in Uranium Mining

Uranium is the source of nuclear power which is widely produced and naturally found. Nuclear power accounts for more than 10% of power production in the world. The uranium deposits are mostly found in Australia, Kazakhstan and Canada. Uranium can be mined using open-pit mining or underground mining. The mined ores are then going through crushing, grinding, leaching, solid separation, liquor extraction, uranium precipitation, solid separation and drying. Table 13 describes the major findings and recommendations from the research conducted related to LCA of this metal. Norgate et al. studied the

**TABLE 13** Summary of the findings and recommendations for LCA of uranium mining.

| Study reference | Key findings | Recommendation |
|---|---|---|
| LCA of nuclear power production (Norgate et al., 2014) | The fuel enrichment stage made the highest contribution of GHG. | New technologies in different stages of the nuclear fuel cycle would be beneficial for reducing environmental burdens. |
| LCA of uranium mining in Canada (Parker et al., 2016) | The energy used to produce electricity and process emissions from nonenergy resources was the largest contributor to environmental emissions. | In future, detailed LCA study covering all the environmental impact categories should be conducted to reduce the environmental burdens and to trade-off among the competing energy products. |
| LCA of global uranium mining (Farjana et al., 2018a,e) | The in situ leaching has a higher impact on all the environmental impact categories except ionising radiation. | In future, further studies should be conducted using the natural gas-based scenario for process and industry-specific datasets. |
| LCA of uranium mining in Australia (Haque and Norgate, 2014) | The good field-related activities and chemical consumption contributed the most for GHG. | Electricity generated from renewable energy resources such as solar energy will be beneficial for reducing impacts from in situ leaching. |

LCA of nuclear power production for uranium while the ore grade was 0.15% $U_3O_8$. The functional unit was 1 MWh of electricity produced from 1 GWe nuclear reactor. The system boundary includes uranium ore mining to electricity production in a plant. Their study found that fuel enrichment stage made the highest contribution of GHG while the total GHG was 34 kg $CO_2$ eq/ MWh. But if the ore grade falls to 0.01%, then the GHG would be 60 kg $CO_2$ eq/MWh (Haque and Norgate, 2014; Norgate et al., 2014). Parker et al. studied the cradle-to-gate uranium mining and milling in Canada. They found that the source of GHG, that is, the energy used to produce electricity and process emissions from nonenergy resources, was the largest contributor to environmental emissions (Parker et al., 2016). Farjana et al. analysed and compared among open-pit mining, underground mining and in situ leaching uranium mining operations. They found that in situ leaching has a higher impact on all the environmental impact categories except ionising radiation. On the other hand, underground mining effects adversely on ionising radiation (Farjana et al., 2018a,e). Haque et al. conducted a comparative LCA study among in situ leaching based production of copper, gold and uranium. Their study was conducted based on Australian Impact method using SimaPro software. They showed that for uranium mining, well field-related activities and chemical consumption contributed the most for GHG. Chemical consumption for leaching and mineral processing contributed about 20% of GHG, while well field-related electricity consumption emits 74% of GHG (Haque and Norgate, 2014).

## LCA in Zinc Mining

Zinc is used for the galvanising process, to protect steel in construction, transportation and products. It is also used as an alloying element with copper. Zinc is produced from ore mining, mineral processing, concentration, roasting, leaching, purification and electrolysis. Table 14 describes the major findings and recommendations from the research conducted related to LCA of this metal. Genderen et al. analysed the cradle-to-gate LCA of zinc concentrate and special high-grade zinc. The geographical coverage considered in that study was global. The system boundary consists of zinc ore mining, concentration, transportation and smelting. Datasets were collected from 24 mines and 18 smelters and analysed using Gabi software. The selected impact categories were GWP, AP, EU (Eutrophication), POCP, ODP and PED. The analysis was done using the CML method, and the functional unit was 1 MT (Mega tonne) of zinc. Their study found that the impacts of zinc production were primarily due to energy consumption in the mining and concentration process. Electricity consumption was huge, which the country and grid mix played a significant role (Van Genderen et al., 2016). Qi et al. conducted LCA of hydrometallurgical zinc production in China, which was based on National inventory datasets. The analysis was done by ReCiPe method, and the

**TABLE 14** Summary of the findings and recommendations for LCA of zinc mining.

| Study reference | Key findings | Recommendation |
|---|---|---|
| LCA of global zinc production (Van Genderen et al., 2016) | The impacts of zinc production were primarily due to energy consumption in the mining and concentration process. | In future, including the unit process details for LCA analysis for concentration and smelting would be beneficial for sustainability. |
| LCA of zinc production in China (Qi et al., 2017) | The overall environmental emissions were due to zinc ore mining and energy consumption in the form of electricity and natural gas. | In future, their study would be beneficial to build national LCI database for zinc production in China. |
| LCA of gold–silver–lead–zinc–copper beneficiation (Farjana et al., 2019b) | Gold–silver beneficiation has the highest impact over the other metals. | Modification in the electricity grid mix to enhance the energy efficiency will be helpful to reduce environmental burdens. |

functional unit was 1 tonne of hydrometallurgical zinc production. Their study reveals that overall environmental emissions were due to zinc ore mining and energy consumption in the form of electricity and natural gas. They also stated that pollutants like copper, zinc, lead, which were emitted to the environment, were responsible for HT, marine aquatic ecotoxicity. Carbon dioxide emission from coal-based electricity generation was dominant for climate change and fossil depletion. Also, the heavy metals from iron ore mining and refining contributed to impacts on human health and water (Qi et al., 2017). Farjana et al. analysed the gold–silver–lead–zinc–copper beneficiation process, which summarises that gold–silver beneficiation has the greatest impact over the other metal in these joint productions which is due to the electricity usage through fossil fuel consumption (Farjana et al., 2019b).

## LCA Studies of Other Metals

Table 15 describes the major findings and recommendations from the research conducted related to LCA of these metals. Cobalt is a valuable metal found in the earth's crust which is widely used in industrial applications. Cobalt mining has a notable impact on human health due to cancer-causing elements which may cause heart disease, vision problem, etc. Farjana et al. conducted the LCA of cobalt extraction process. According to their study, cobalt extraction is harmful to eutrophication and global warming. Cobalt extraction requires a

**TABLE 15** Summary of the findings and recommendations for LCA of different metals.

| Study reference | Key findings | Recommendation |
|---|---|---|
| LCA of the cobalt extraction process (Farjana et al., 2019c) | Eutrophication and global warming impacts are higher due to electricity consumption. | Altering the electricity generation resources within the same grid mix network will be beneficial for sustainability. |
| LCA of non-Chinese cemented carbide production (Furberg et al., 2019) | The impacts were due to elements like kerosene, tailings, water and electricity. | In future, improvement in LCI datasets is required to get rid of uncertainty in environmental sustainability. |
| LCA of global manganese production (Westfall et al., 2016) | The electricity demand, fuel consumption during smelting were the primary contributors for impact. | In future, the LCI datasets and LCA results can be integrated into cost and process optimization |
| LCA of magnesium oxide production (Ruan and Unluer, 2016) | MgO has a lower impact on the ecosystem and resources but a larger impact on human health. | An extensive database with detailed LCI should be built to conduct detailed LCA analysis. Moreover, the reduction of carbon dioxide emissions should be achieved. |
| LCA of silica sand (Grbeš, 2016) | Fossil fuel-based processes are found as most impactful over others. | They recommended to focus on petrol-based processes and lowering fossil fuel use. |
| LCA of titanium oxides in Australia (Farjana et al., 2018c) | Rutile had a significant environmental impact than for ilmenite due to higher energy consumption and electricity use. | Electricity grid mix has a significant impact on the environmental impact analysis results, which would be analysed extensively in future. |
| LCA of gold–silver–lead–zinc–copper beneficiation (Farjana et al., 2019b) | Gold–silver beneficiation has the highest impact over the other metals. | Modification in the electricity grid mix to enhance the energy efficiency will be helpful to reduce environmental burdens. |
| LCA of gold–silver refining operations (Farjana et al., 2019d) | Gold–silver refining from the couple production of gold–silver has more environmental burdens associated with gold–silver–lead–zinc–copper couple production. | Altering the stainless steel alloying properties would be beneficial for sustainability. |

large amount of electricity which is detrimental to global warming and also is the blasting (Farjana et al., 2019c). Cemented carbide has higher hardness and higher corrosion resistance, mostly used for drilling tools and cutting tools. China is the leading producer of cemented carbide. The cemented carbide ore is mined from extraction, crushing, milling, gravity method grinding, sulphide flotation and roasting. In the hydrometallurgy stage, the cemented carbide ore is digested, filtrated, precipitated, extracted using solvent and finally crystalised. In the pyrometallurgy stage, the ore goes through calcination, hydrogen reduction and carburisation. In the powder metallurgy stage, the ore goes through powder milling, granulation and sintering. Furberg et al. conducted a cradle-to-gate LCA of cemented carbide production with cobalt, while the geographic location was non-Chinese (Canada and United States). Their study stated that impacts were due to elements like kerosene, tailings, water and electricity. The highest impacts were on the category of TAP (terrestrial acidification), ODP (ozone depletion), FEU (freshwater eutrophication). And the lowest impact was on CC (climate change), PCOF (photochemical oxidant formation) and WD (water depletion) (Furberg et al., 2019). Manganese is an essential element for batteries, fertilisers and chemicals. Manganese is a widely used alloying element that comes in conjunction to make ferroalloys. The manganese alloy is produced using mineral extraction, hauling, ore preparation and beneficiation, sintering and transportation, smelting, crushing, screening and refining. Westfall et al. conducted an LCA study based on manganese alloys, where datasets were collected from 16 ore and alloy producers. The authors have conducted a cradle-to-gate LCA of silicomanganese, ferromanganese and refined ferromanganese. The impact categories considered were GWP, AP, POCP, water and waste. The analysis was done using CML 2001 method. According to their analysis, electricity demand, fuel consumption during smelting was the primary contributors for impact (Westfall et al., 2016). Magnesium oxide cement is widely produced in China, North Korea, Turkey, Russia and Australia. The magnesium oxide is produced from raw material acquisition, crushing, vertical shaft kiln, precipitation tank, screening, crushing, grinding and packaging. Ruan et al. analysed the LCA of magnesium oxide, where the functional unit was 1 tonne. They showed that MgO has a lower impact on the ecosystem and resources but a larger impact on human health. The analysis was done using EcoIndicator 99 method. They considered five different case scenarios based on fossil fuel and raw material consumption (Ruan and Unluer, 2016). Silver metal is most widely used for industrial purposes or for making jewellery. There are very few studies which addressed the environmental impact of silver mining processes. Farjana et al. analysed the environmental burdens associated with gold—silver—lead—zinc—copper beneficiation process (Farjana et al., 2019b). In another study, they analysed the environmental impacts of gold—silver refining operations (Farjana et al., 2019d). They found that silver beneficiation and refining have the least environmental impacts than gold mining processes as they consume the least amount of electricity. However, there are some impacts on eutrophication, global warming and ecotoxicity (Farjana et al., 2019b,e). Titanium oxides are widely used for making high-performance metal parts, artificial body parts and engine elements. Ilmenite and rutile are the generally found form of titanium oxides. Ilmenite and rutile are extracted from mining site using heavy mineral concentration, rare-earth drum separation, electrostatic separation circuit and gravity separation circuit. Farjana et al. conducted a comparative LCA analysis of cradle-to-gate titanium oxides production. Ilmenite and rutile were considered where the geographic region considered was for Australia. The datasets were collected from the AusLCI database and SimaPro software. The study revealed that rutile had a significant environmental impact than for ilmenite due to higher energy consumption and electricity use. The GHG was 0.295 kg $CO_2$ eq/kg of ilmenite production and 1.535 kg $CO_2$ eq/kg of rutile production (Farjana et al., 2018c).

## Discussion

This chapter discusses the key aspects of the research articles on LCA of metal and mining industries. Analysis results can be discussed based on a combination of four different perspectives — choice of LCA methods to be used for accessing the impacts of the mining processes, mining technologies, environmental impacts and energy consumption and integration. The two most widely used method for LCA analysis is ReCiPe and ILCD, although it is hard to compare among the papers for the same metals due to their processes under consideration, geographic location, inputs and outputs of production and production technologies. Figs 2 and 3 show the comparative analysis results for few studies similar in process consideration irrespective of their data source and geography. Fig. 2 is focussed on the comparison of metals analysed using ReCiPe midpoint indicator-based methods while the metals are gold, cemented carbide, silica, steel and zinc. According to the results presented here, it is evident that gold mining has the highest impact over the other metals, followed by zinc and steel. The detailed results are presented in Table 16. Similarly, Fig. 3 describes the comparative analysis results for the LCA studies based on the ILCD method, which are only focussed on the refining processes. The metals are aluminium, gold and silver refining operations. As shown in the figure, gold refining has the highest impact, followed by the aluminium refining operations. The detailed results are described in Table 17 below.

According to the discussion on the LCA results presented in Table 18, which describes the key findings of these research papers on LCA based on metal type, it is quite evident that the common reason for largest environmental impact from mining is due to energy generation, fossil fuel consumption for energy and process heat generation. For aluminium, smelting and refining stage made the greatest contribution due to energy consumption and

## Comparative Results using ReCiPe method

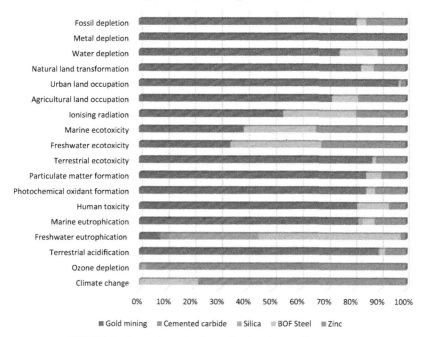

**FIGURE 2**   Comparison of LCA studies based on ReCiPe method.

## Comparative Results using ILCD method

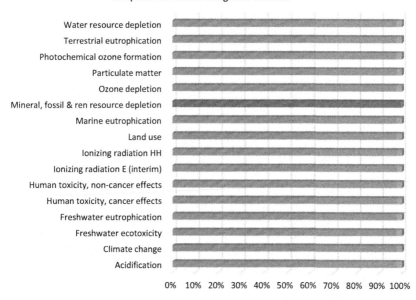

**FIGURE 3**   Comparison of LCA studies based on the ILCD method.

**TABLE 16** Detailed analysis results for studies utilising ReCiPe method.

| Impact categories | Unit | Gold (Chen et al., 2018) | Cemented carbide (Furberg et al., 2019) | Silica (Grbeš, 2016) | BOF Steel (Burchart-Korol, 2013) | Zinc (Qi et al., 2017) |
|---|---|---|---|---|---|---|
| Climate change | kg $CO_2$ eq | $5.55 \times 10^4$ | 14 | 29.6 | 1703 | 6.12E+03 |
| Ozone depletion | kg CFC-11 eq | $1.15 \times 10^{-4}$ | 2.60E-06 | 4.48E-06 | | 2.76E-04 |
| Terrestrial acidification | kg $SO_2$ eq | 207.25 | 0.58 | 0.725 | 4.81 | 18.28 |
| Freshwater eutrophication | kg P eq | 0.12 | 0.054 | 0.507 | 0.81 | 0.03 |
| Marine eutrophication | kg N eq | 5.55 | | 0.113 | 0.3 | 0.81 |
| Human toxicity | kg 1,4-DB eq | 4.37E+03 | | 2.27 | 643 | 348.89 |
| Photochemical oxidant formation | kg NMVOC | 130.84 | 0.049 | 0.277 | 4.89 | 18.15 |
| Particulate matter formation | kg $PM_{10}$ eq | 67.6 | | 6.78E-04 | 4.61 | 7.47 |
| Terrestrial ecotoxicity | kg 1,4-DB eq | 11.5 | | 1.75E-02 | 0.17 | 1.51 |

| Freshwater ecotoxicity | kg 1,4-DB eq | 13.06 | | 8.17E-04 | 12.77 | 12.19 |
|---|---|---|---|---|---|---|
| Marine ecotoxicity | kg 1,4-DB eq | 19.61 | | 2.27E-02 | 13.32 | 17 |
| Ionising radiation | kBq U$^{235}$ eq | 163.67 | | 2.13E-02 | 82.83 | 57.54 |
| Agricultural land occupation | m$^2$a | 330.41 | | 4.16E-03 | 45.55 | 83.42 |
| Urban land occupation | m$^2$a | 1.69E+03 | | 2.90E-01 | 12.21 | 34.04 |
| Natural land transformation | m$^2$ | 3.57 | | 5.54E-03 | 0.2 | 0.53 |
| Water depletion | m$^3$ | 461.25 | 0.24 | 1.43E+00 | 87.44 | 66.2 |
| Metal depletion | kg Fe eq | 2.16E+10 | | 4.57E-02 | 850 | 3.58E+03 |
| Fossil depletion | kg oil eq | 9.98E+03 | | 1.01E-01 | 429 | 1.86E+03 |

**TABLE 17** Detailed analysis results for studies utilising the ILCD method.

| Impact category | Unit | Aluminium (Farjana et al., 2019a) | Gold (Farjana et al., 2019e) | Silver (Farjana et al., 2019e) |
|---|---|---|---|---|
| Acidification | molc $H^+$ eq | 2.60E-03 | 280.98 | 6.79 |
| Climate change | kg $CO_2$ eq | 0.27 | 3.40E+04 | 815.44 |
| Freshwater ecotoxicity | CTUe | 0.36 | 1.30E+04 | 330.61 |
| Freshwater eutrophication | kg P eq | 0 | 65.45 | 1.58 |
| Human toxicity, cancer effects | CTUh | 1.05E-09 | 6.00E-04 | 1.45E-05 |
| Human toxicity, noncancer effects | CTUh | 1.86E-08 | 1.00E-03 | 2.42E-05 |
| Ionising radiation E (interim) | CTUe | N/A | 4.20E-04 | 1.02E-05 |
| Ionising radiation HH | kBq $U^{235}$ eq | N/A | 47.15 | 1.14 |
| Land use | kg C deficit | N/A | 1.60E+04 | 389.39 |
| Marine eutrophication | kg N eq | 2.18E-04 | 100.14 | 2.42 |
| Mineral, fossil and renewable resource depletion | kg Sb eq | 4.61E-08 | 1.02E-10 | 2.45E-12 |
| Ozone depletion | kg CFC-11 eq | 3.19E-12 | 2.20E-03 | 5.31E-05 |
| Particulate matter | kg PM2.5 eq | 1.07E-04 | 28.56 | 0.689 |
| Photochemical ozone formation | kg NMVOC eq | 7.60E-04 | 295.84 | 7.14 |
| Terrestrial eutrophication | molc N eq | 2.30E-03 | 1.20E+03 | 29.95 |
| Water resource depletion | $M^3$ water eq | 0 | 225.67 | 5.45 |

**TABLE 18** Key factors of environmental impacts from metal mining industries.

| Metal | Highest impactful category | Impactful process |
|---|---|---|
| Aluminium (Nunez and Jones, 2016; Paraskevas et al., 2016; Tan and Khoo, 2005) | GWP due to smelting and refining | Smelting and refining, electricity consumption and direct emissions. |
| Cobalt (Farjana et al., 2019c) | GWP, EU | Extraction due to electricity consumption and blasting. |
| Cemented carbide (Furberg et al., 2019) | TE, OD, FWE | Mining, hydrometallurgy and powder metallurgy phases. Due to kerosene, sulphidic tailing, water use, and electricity use. |
| Coal (Adiansyah et al., 2017; Burchart-Korol et al., 2016; Guimarães da Silva et al., 2018; Zhang et al., 2018) | Dust, GWP, AP | Mining due to electricity usage, diesel oil consumption, fossil fuel, use of steel and heat. |
| Copper (Ekman Nilsson et al., 2017; Haque and Norgate, 2014; Memary et al., 2012; Norgate, 2001; Northey et al., 2013) | GWP | Solvent extraction and electrowinning, leaching due to electricity usage, fossil fuel, sulphuric acid. |
| Ferroalloy (Bartzas and Komnitsas, 2015; Haque and Norgate, 2013) | GWP | Smelting and refining due to electricity usage, fossil fuel. |
| Gold (Chen et al., 2018; Farjana et al., 2019b,d; Haque and Norgate, 2014; Norgate and Haque, 2012) | GWP, MD | Mining and comminution, well field-related activities due to electricity usage, diesel oil consumption and direct emissions. |
| Iron (Ferreira and Leite, 2015; Gan and Griffin, 2018; Haque and Norgate, 2015; Norgate and Haque, 2010) | GHG, human health | Agglomeration, loading and haulage, ore processing, vegetation and soil removal due to electricity usage, emission of inorganic materials, use of grinding media. |
| Manganese (Westfall et al., 2016) | GHG | Smelting due to electricity usage, fossil fuel (coal and coke). |

*Continued*

**TABLE 18** Key factors of environmental impacts from metal mining industries.—cont'd

| Metal | Highest impactful category | Impactful process |
|---|---|---|
| Magnesium oxide (Ruan and Unluer, 2016) | Human health | Calcination due to coal usage. |
| Nickel (Khoo et al., 2017; Mistry et al., 2016) | GHG | Primary extraction, nickel reduction and smelting due to electricity usage, coal consumption to produce steam. |
| A rare-earth element (Lima et al., 2018; Zaimes et al., 2015) | ODP, carcinogens | Oxide separation step due to consumption of hydrochloric acid and ammonium oxides. |
| Silver (Farjana et al., 2019b,d) | GWP, EU | Stainless steel and electricity used in beneficiation and refining processes. |
| Steel (Burchart-Korol, 2013; Norgate et al., 2007; Renzulli et al., 2016) | GHG, dust | Pig iron in the blast furnace and coke oven operations, the iron ore sintering process due to electricity usage, fossil fuel. |
| Titanium oxides (Farjana et al., 2018c) | GHG | Mining due to electricity usage. |
| Uranium (Farjana et al., 2018e; Haque and Norgate, 2014; Norgate et al., 2014; Parker et al., 2016) | GHG | Fuel enrichment, well field-related activities, leaching due to electricity usage, use of chemicals and direct emissions. |
| Zinc (Qi et al., 2017; Van Genderen et al., 2016) | GWP | Mining and concentration due to electricity consumption and fossil fuel usage. |

direct emissions to the environment. From the cemented carbide production, the mining, hydrometallurgy and powder metallurgy stages made the greatest contribution due to kerosene use, sulphidic tailing, water and electricity.

In the case of coal mining, still electricity use, fossil fuel use for heat generation creates greater environmental impacts generated from the mining stage. Among the copper mining steps, solvent extraction and electrowinning in hydrometallurgy causes greater environmental impact rather than for pyrometallurgy, due to electricity, fossil fuel use for heat generation and steel use. Gold mining is impactful in mining and comminution stage, due to well field-related activities. Agglomeration, loading, haulage and ore processing in iron mining is harmful to the environment due to the emission of inorganic chemical materials, use of electricity and use of grinding media. From magnesium oxide production, the calcination stage is highly impactful towards the environment due to fossil fuel usage in the form of coal. For nickel production, primary extraction, nickel reduction and smelting are impactful due to electricity usage and coal consumption to produce steam. The oxide separation stage in rare-earth mining step consumes a huge amount of hydrochloric acid and ammonium oxides which is harmful. Pig iron in a blast furnace and iron ore sintering process in steel and stainless steel production causes a greater impact due to electricity and fossil fuel usage. The fuel enrichment stage, well field-related stages and the mining stage cause the greatest environmental impact of uranium mining due to electricity usage, use of chemicals and indirect emissions. In summary, mining and extraction stages made the greatest contribution in cemented carbide, coal, gold, titanium oxides and zinc production. Smelting and refining steps cause a higher impact for aluminium, ferroalloy, nickel and manganese mining which are due to the significant amount of electricity consumed in the smelting operation.

From the discussion presented above, it is quite evident that most of the environmental emissions from mining industries are due to electricity consumption and fossil fuel usage, whatever the process is. Fossil fuel can be of the form of diesel oil, natural gas, heavy fuel oil and residual fuel oil. These fossil fuels can be used for electricity production, process heat generation and mechanical energy generation. Among these, fossil fuel use can be reduced or replaced by integrating renewable energy resources. Renewable energy can be used as an alternative fuel to generate electricity or to produce process heat. Several LCA studies assessed the environmental impact scenario with the integration of renewable energy resources in mining industries. There could be the replacement by different types of renewables either partially or completely. Another major reason is the electricity grid mix, which varies from one country to another. The grid mix is a combination of fuel sources used to generate electricity. Among the three major components in LCA, electricity mix, technology mix and heat mix-electricity mix and heat mix both are related to fossil fuel. Between them, a most adverse effect is from electricity use which impacts on GWP and AP. It also affects PED and GER. Replacing coal with natural gas may also be a solution to reduce GWP with same GER, but natural gas has more ionising radiation impact on human health. However, it is possible but dependent on the availability of the renewables at the mining site to integrate with the power generation system. This replacement is time-consuming and involves huge capital cost. Another important point is electricity generation efficiencies which could be improved after involving modern technologies.

Another major part is the integration of renewable resources to generate process heat in mining. Regarding solar energy, temperatures below 150°C could easily be achieved by non-CST (nonconcentrated solar thermal) technologies. Temperature ranging between 150 and 400°C is possible to generate with primary CST (concentrated solar thermal) technologies. There are few industries in the world where already process heat integration is operating with solar technologies. Few mines in Chile are utilising solar process heat for electrowinning operations. South Africa is using solar process heat for mining and cleaning operations. Cyprus, Austria, India and Germany are also using solar process heat for different low-temperature mining operations which all operate below 100°C. The availability of solar energy also varies depending on the latitude and longitude, irradiation and many other factors. Not all miner countries could have this renewable source of energy (Farjana et al., 2018d,f).

Other than energy integration, there are some technological factors which affect the environmental impacts of mining processes, the most important factor, which was discussed in almost every paper, is that ore grade effects largely on environmental burdens associated with mining. The deeper would the mining goes, the greater machinery requirements and the energy consumption would be. The more we are approaching towards development, more mining of the metals is required. Another important factor is the use of inorganic chemicals for chemical extraction, acid leaching and electrowinning processes, which is more troublesome for environmental sustainability. Mining tailing management is another critical issue, which causes a significant load on the environment impacting human health, ecosystems and resources.

## Conclusion

In conclusion, this chapter presents a comprehensive scenario of all the existing studies reported in the literature on the life cycle environmental impact analysis of metal mining industries. Sixteen metal mining industries are identified which have significant research output that quantified the environmental effects based on LCA methods. A comprehensive compilation of the published work is being carried out, analysed and compared the impactful mining materials, processes which are energy-intensive, thus harmful for global environmental sustainability. According to the findings presented in this chapter, mining and extraction stages made the notable contribution in cemented carbide, coal, gold, titanium oxides and zinc production. Also, smelting and refining steps cause a considerable amount of impact for aluminium, ferroalloy, nickel and manganese mining which are due to the significant amount of electricity consumed in the smelting operation. This chapter also summarises the key methods of LCA which were utilised for mining industries. Limitations of these studies and future recommendation to improve environmental sustainability are also presented. The most important challenge is the replacement of energy generation resources by renewables, which could significantly reduce the gross energy requirements and GWP, to achieve the sustainability goals for 2050.

In summary, there are some limitations and future recommendations focussed on LCA techniques and methods as follows:

**a.** Development of global LCA method which would be applicable irrespective of the geographic region of the mining company.
**b.** Development of aggregated allocation technique for coproduction of metals, which would be free from technological and economic differences of metals.
**c.** Analysis of complete LCA for each metal industries, as most of them, lack full LCA study, due to the research focus and availability of dataset.
**d.** Development of a specific database focussed on mining industries LCI datasets, which should contain inventory datasets for every type of mining technology per metal. According to the analysis results presented in this study, it is evident that the mining industry is considerably affecting GWP and human health (carcinogenic and noncarcinogenic).

The following measures could be adopted for a significant reduction in the emission during the mining processes:

**a.** Replacement of electricity generation sources by renewables is of paramount importance. The choice of renewable energy resource would be dependent on the geographic location of the mine, availability of energy resources and the capital cost of replacement.
**b.** Process heat generation sources could be altered by renewables. Mining processes should be classified as in terms of heating needs as low-temperature, medium temperature and high-temperature process heat applications. Following that, the feasibility of process heat integration should be assessed approaching towards the practical integration.
**c.** Sourcing high-grade mining ore would be beneficial for the environment, which will consume less energy in the form of electricity and heat and less equipment to extract mined ore.
**d.** Decreasing the use of inorganic chemicals for leaching or increasing the efficiency of leaching processes would be beneficial to reduce the human health impacts, both carcinogenic and noncarcinogenic.

# References

Acero, A.A.P., Rodríguez, C., Ciroth, A., 2015. LCIA Methods Impact Assessment Methods in Life Cycle Assessment and Their Impact Categories 1−22.

Adiansyah, J.S., Haque, N., Rosano, M., Biswas, W., 2017. Application of a life cycle assessment to compare environmental performance in coal mine tailings management. J. Environ. Manag. 199, 181−191. https://doi.org/10.1016/j.jenvman.2017.05.050.

Althaus, H.J., Classen, M., 2005. Life cycle inventories of metals and methodological aspects of inventorying material resources in ecoinvent. Int. J. Life Cycle Assess. 10, 43−49. https://doi.org/10.1065/lca2004.11.181.5.

Althaus, H., Chudacoff, M., Hischier, R., Jungbluth, N., Osses, M., Primas, A., Hellweg, S., 2007. Life Cycle Inventories of Chemicals, pp. 1−957. Ecoinvent data v2.0, Swiss Centre for Life Cycle Inventories.

Bartzas, G., Komnitsas, K., 2015. Life cycle assessment of ferronickel production in Greece. Resour. Conserv. Recycl. 105, 113−122. https://doi.org/10.1016/j.resconrec.2015.10.016.

Burchart-Korol, D., 2013. Life cycle assessment of steel production in Poland: a case study. J. Clean. Prod. 54, 235−243. https://doi.org/10.1016/j.jclepro.2013.04.031.

Burchart-Korol, D., Fugiel, A., Czaplicka-Kolarz, K., Turek, M., 2016. Model of environmental life cycle assessment for coal mining operations. Sci. Total Environ. 562, 61−72. https://doi.org/10.1016/j.scitotenv.2016.03.202.

Chen, W., Geng, Y., Hong, J., Dong, H., Cui, X., Sun, M., Zhang, Q., 2018. Life cycle assessment of gold production in China. J. Clean. Prod. 179, 143−150. https://doi.org/10.1016/j.jclepro.2018.01.114.

Cortez-Lugo, M., Riojas-Rodríguez, H., Moreno-Macías, H., Montes, S., Rodríguez-Agudelo, Y., Hernández-Bonilla, D., Catalán-Vázquez, M., Díaz-Godoy, R., Rodríguez-Dozal, S., 2018. Evaluation of the effect of an environmental management program on exposure to manganese in a mining zone in Mexico. Neurotoxicology 64, 142−151. https://doi.org/10.1016/j.neuro.2017.08.014.

Curran, M.A., 2012. Life Cycle Assessment Handbook. https://doi.org/10.1002/9781118528372.

Ekman Nilsson, A., Macias Aragonés, M., Arroyo Torralvo, F., Dunon, V., Angel, H., Komnitsas, K., Willquist, K., 2017. A review of the carbon footprint of Cu and Zn production from primary and secondary sources. Minerals 7, 168. https://doi.org/10.3390/min7090168.

EPA, 2018. The emissions & generation resource integrated database (eGRID) technical support document. US Environ. Prot. Agency 106.

European Commission − Joint Research Centre − Institute for Environment and Sustainability, 2010. International Reference Life Cycle Data System (ILCD) Handbook − General Guide for Life Cycle Assessment − Detailed Guidance, Constraints. https://doi.org/10.2788/38479.

Farjana, S.H., Huda, N., Mahmud, M.A.P., 2018a. Life-Cycle environmental impact assessment of mineral industries. In: IOP Conference Series: Materials Science and Engineering. https://doi.org/10.1088/1757-899X/351/1/012016.

Farjana, S.H., Huda, N., Mahmud, M.A.P., Lang, C., 2018b. Life-Cycle Environmental Impact Assessment of Mineral Industries Life-Cycle Environmental Impact Assessment of Mineral Industries. https://doi.org/10.1088/1757-899X/351/1/012016.

Farjana, S.H., Huda, N., Mahmud, M.A.P., Lang, C., 2018c. Towards sustainable $TiO_2$ production: an investigation of environmental impacts of ilmenite and rutile processing routes in Australia. J. Clean. Prod. 196, 1016−1025. https://doi.org/10.1016/j.jclepro.2018.06.156.

Farjana, S.H., Huda, N., Mahmud, M.A.P., Saidur, R., 2018d. Solar process heat in industrial systems — a global review. Renew. Sustain. Energy Rev. 82 https://doi.org/10.1016/j.rser.2017.08.065.

Farjana, S.H., Huda, N., Mahmud, M.A.P., Lang, C., 2018e. Comparative life-cycle assessment of uranium extraction processes in Australia. J. Clean. Prod. 202, 666—683. https://doi.org/10.1016/j.jclepro.2018.08.105.

Farjana, S.H., Huda, N., Mahmud, M.A.P., Saidur, R., 2018f. Solar industrial process heating systems in operation — current SHIP plants and future prospects in Australia. Renew. Sustain. Energy Rev. 91 https://doi.org/10.1016/j.rser.2018.03.105.

Farjana, S.H., Huda, N., Mahmud, M.A.P., 2019c. Life cycle assessment of cobalt extraction process. J. Sustain. Min. 18 (3), 150—161. https://doi.org/10.1016/j.jsm.2019.03.002.

Farjana, S.H., Huda, N., Mahmud, M.A.P., 2019a. Impacts of aluminum production: a cradle to gate investigation using life-cycle assessment. Sci. Total Environ. 663, 958—970. https://doi.org/10.1016/j.scitotenv.2019.01.400.

Farjana, S.H., Huda, N., Mahmud, M.A.P., 2019b. Life cycle analysis of copper-gold-lead-silver-zinc beneficiation process. Sci. Total Environ. 659, 41—52. https://doi.org/10.1016/j.scitotenv.2018.12.318.

Farjana, S.H., Huda, N., Mahmud, M.A.P., Lang, C., 2019d. Impact analysis of gold-silver refining processes through life-cycle assessment. J. Clean. Prod. 228, 867—881. https://doi.org/10.1016/j.jclepro.2019.04.166.

Farjana, S.H., Huda, N., Mahmud, M.A.P., Lang, C., 2019e. Impact analysis of gold-silver refining processes through life-cycle assessment. J. Clean. Prod. 228 https://doi.org/10.1016/j.jclepro.2019.04.166.

Ferreira, H., Leite, M.G.P., 2015. A life cycle assessment study of iron ore mining. J. Clean. Prod. 108, 1081—1091. https://doi.org/10.1016/j.jclepro.2015.05.140.

Fogler, S., Timmons, D., 1998. An Overview of the ISO 14040 Life Cycle Assessment Approach and an Industrial Case Study. Argentum V.

Fugiel, A., Burchart-Korol, D., Czaplicka-Kolarz, K., Smoliński, A., 2017. Environmental impact and damage categories caused by air pollution emissions from mining and quarrying sectors of European countries. J. Clean. Prod. 143, 159—168. https://doi.org/10.1016/j.jclepro.2016.12.136.

Furberg, A., Arvidsson, R., Molander, S., 2019. Environmental life cycle assessment of cemented carbide (WC-Co) production. J. Clean. Prod. 209, 1126—1138. https://doi.org/10.1016/j.jclepro.2018.10.272.

Gan, Y., Griffin, W.M., 2018. Analysis of life-cycle GHG emissions for iron ore mining and processing in China—uncertainty and trends. Resour. Pol. 58, 90—96. https://doi.org/10.1016/j.resourpol.2018.03.015.

Goedkoop, M., Oele, M., Vieira, M., Leijting, J., Ponsioen, T., Meijer, E., 2014. SimaPro Tutorial, vol. 89. https://doi.org/10.1142/S0218625X03005293.

Gorman, M.R., Dzombak, D.A., 2018. A review of sustainable mining and resource management: transitioning from the life cycle of the mine to the life cycle of the mineral. Resour. Conserv. Recycl. 137, 281—291. https://doi.org/10.1016/j.resconrec.2018.06.001.

Grande, C.A., Blom, R., Spjelkavik, A., Moreau, V., Payet, J., 2017. Life-cycle assessment as a tool for eco-design of metal-organic frameworks (MOFs). Sustain. Mater. Technol. 14, 11—18. https://doi.org/10.1016/j.susmat.2017.10.002.

Grbeš, A., 2016. A life cycle assessment of silica sand: comparing the beneficiation processes. Sustainability 8, 1—9. https://doi.org/10.3390/su8010011.

Guimarães da Silva, M., Costa Muniz, A.R., Hoffmann, R., Luz Lisbôa, A.C., 2018. Impact of greenhouse gases on surface coal mining in Brazil. J. Clean. Prod. 193, 206–216. https://doi.org/10.1016/j.jclepro.2018.05.076.

Haque, N., Norgate, T., 2013. Estimation of greenhouse gas emissions from ferroalloy production using life cycle assessment with particular reference to Australia. J. Clean. Prod. 39, 220–230. https://doi.org/10.1016/j.jclepro.2012.08.010.

Haque, N., Norgate, T., 2014. The greenhouse gas footprint of in-situ leaching of uranium, gold and copper in Australia. J. Clean. Prod. 84, 382–390. https://doi.org/10.1016/j.jclepro.2013.09.033.

Haque, N., Norgate, T., 2015. Life Cycle Assessment of Iron Ore Mining and Processing, Iron Ore: Mineralogy, Processing and Environmental Sustainability. Elsevier Ltd. https://doi.org/10.1016/B978-1-78242-156-6.00020-4.

Haque, N., Norgate, T., Northey, S., 2014. Life cycle based greenhouse gas footprints of metal production with recycling scenarios. TMS Annu. Meet. 113–120. https://doi.org/10.1002/9781118889664.ch14.

Hischier, R., Weidema, B., Althaus, H.-J., Bauer, C., Doka, G., Dones, R., Frischknecht, R., Hellweg, S., Humbert, S., Jungbluth, N., Köllner, T., Loerincik, Y., Margni, M., Nemecek, T., 2010. Implementation of Life Cycle Impact Assessment Methods Data v2.2 (2010). Ecoinvent Rep. No. 3 176.

JRC European commission, 2011. ILCD Handbook: Recommendations for Life Cycle Impact Assessment in the European Context. https://doi.org/10.2788/33030. Vasa.

Khoo, J.Z., Haque, N., Woodbridge, G., McDonald, R., Bhattacharya, S., 2017. A life cycle assessment of a new laterite processing technology. J. Clean. Prod. 142, 1765–1777. https://doi.org/10.1016/j.jclepro.2016.11.111.

Lima, F.M., Lovon-Canchumani, G.A., Sampaio, M., Tarazona-Alvarado, L.M., 2018. Life cycle assessment of the production of rare earth oxides from a Brazilian ore. Procedia CIRP 69, 481–486. https://doi.org/10.1016/j.procir.2017.11.066.

Lodhia, S., Hess, N., 2014. Sustainability accounting and reporting in the mining industry: current literature and directions for future research. J. Clean. Prod. 84, 43–50. https://doi.org/10.1016/j.jclepro.2014.08.094.

Long, K.R., DeYoung, J.H., Ludington, S.D., 1998. Database of significant deposits of gold, silver, copper, lead, and zinc in the United States. Part A: database description and analysis. Open-File Report 98-206A. US Geol. Surv. 1–60.

Ma, X., Ye, L., Qi, C., Yang, D., Shen, X., Hong, J., 2018. Life cycle assessment and water footprint evaluation of crude steel production: a case study in China. J. Environ. Manag. 224, 10–18. https://doi.org/10.1016/j.jenvman.2018.07.027.

Mahmud, M.A.P., Huda, N., Farjana, S.H., 2018a. Environmental Profile Evaluations of Piezo-electric Polymers Using Life Cycle Assessment Environmental Profile Evaluations of Piezoelectric Polymers Using Life Cycle Assessment.

Mahmud, M.A.P., Huda, N., Farjana, S.H., Lang, C., 2018b. Environmental life-cycle assessment and techno-economic analysis of photovoltaic (PV) and photovoltaic/thermal (PV/T) systems. IEEE Int. Conf. Environ. Electr. Eng. 2018 IEEE Ind. Commer. Power Syst. Eur. (EEEIC/I&CPS Eur.) 1–5.

Mahmud, M., Huda, N., Farjana, S., Lang, C., Mahmud, M.A.P., Huda, N., Farjana, S.H., Lang, C., 2018c. Environmental impacts of solar-photovoltaic and solar-thermal systems with life-cycle assessment. Energies 11. https://doi.org/10.3390/EN11092346, 2346 11, 2346.

Mahmud, M.A.P., Huda, N., Farjana, S.H., Lang, C., 2018d. Environmental sustainability assessment of hydropower plant in Europe using life cycle assessment. In: IOP Conference Series: Materials Science and Engineering. https://doi.org/10.1088/1757-899X/351/1/012006.

Mahmud, M.A.P., Huda, N., Farjana, S.H., Lang, C., 2019. A strategic impact assessment of hydropower plants in alpine and non-alpine areas of Europe. Appl. Energy 250, 198−214. https://doi.org/10.1016/j.apenergy.2019.05.007.

Marguerite, R., Tim, G., Maartje, S., James, L., Brad, R., Fabiano, X., Jonas, B., Annette, C., Joe, J., 2015. Best practice guide for life cycle impact assessment (LCIA) in Australia ALCAS impact assessment committee. ALCAS Impact Assess. Comm. http://www.auslci.com.au/Documents/Best_Practice_Guide_V2_Draft_for_Consultation.pdf.

Memary, R., Giurco, D., Mudd, G., Mason, L., 2012. Life cycle assessment: a time-series analysis of copper. J. Clean. Prod. 33, 97−108. https://doi.org/10.1016/j.jclepro.2012.04.025.

Menoufi, K.A.I., 2011. Life Cycle Analysis and Life Cycle Impact Assessment Methodologies: State of the Art (Master thesis), pp. 1−84.

Mistry, M., Gediga, J., Boonzaier, S., 2016. Life cycle assessment of nickel products. Int. J. Life Cycle Assess. 21, 1559−1572. https://doi.org/10.1007/s11367-016-1085-x.

Mudd, G.M., 2009. The Sustainability of Mining in Australia: Key Production Trends and Their Environmental Implications for the Future. ISBN: 978-0-9803199-4-1.

Mutchek, M., Cooney, G., Pickenpaugh, G., Marriott, J., Skone, T., 2016. Understanding the contribution of mining and transportation to the total life cycle impacts of coal exported from the United States. Energies 9, 559. https://doi.org/10.3390/en9070559.

Norgate, T.E., 2001. A Comparative Life Cycle Assessment of Copper Production Processes.

Norgate, T., Haque, N., 2010. Energy and greenhouse gas impacts of mining and mineral processing operations. J. Clean. Prod. 18, 266−274. https://doi.org/10.1016/j.jclepro.2009.09.020.

Norgate, T., Haque, N., 2012. Using life cycle assessment to evaluate some environmental impacts of gold production. J. Clean. Prod. 29 (30), 53−63. https://doi.org/10.1016/j.jclepro.2012.01.042.

Norgate, T.E., Jahanshahi, S., Rankin, W.J., 2007. Assessing the environmental impact of metal production processes. J. Clean. Prod. 15, 838−848. https://doi.org/10.1016/j.jclepro.2006.06.018.

Norgate, T., Haque, N., Koltun, P., 2014. The impact of uranium ore grade on the greenhouse gas footprint of nuclear power. J. Clean. Prod. 84, 360−367. https://doi.org/10.1016/j.jclepro.2013.11.034.

Northey, S., Haque, N., Mudd, G., 2013. Using sustainability reporting to assess the environmental footprint of copper mining. J. Clean. Prod. 40, 118−128. https://doi.org/10.1016/j.jclepro.2012.09.027.

Nunez, P., Jones, S., 2016. Cradle to gate: life cycle impact of primary aluminium production. Int. J. Life Cycle Assess. 21, 1594−1604. https://doi.org/10.1007/s11367-015-1003-7.

Paraskevas, D., Kellens, K., Van De Voorde, A., Dewulf, W., Duflou, J.R., 2016. Environmental impact analysis of primary aluminium production at country level. Procedia CIRP 40, 209−213. https://doi.org/10.1016/j.procir.2016.01.104.

Parker, D.J., MNaughton, C.S., Sparks, G.A., 2016. Life cycle greenhouse gas emissions from uranium mining and milling in Canada. Environ. Sci. Technol. 50, 9746−9753. https://doi.org/10.1021/acs.est.5b06072.

PRé, 2018. SimaPro Database Manual Methods Library Colophon Title: SimaPro Database Manual Methods Library. https://doi.org/10.1017/CBO9781107415324.004.

Qi, C., Ye, L., Ma, X., Yang, D., Hong, J., 2017. Life cycle assessment of the hydrometallurgical zinc production chain in China. J. Clean. Prod. 156, 451−458. https://doi.org/10.1016/j.jclepro.2017.04.084.

Ranängen, H., Lindman, Å., 2017. A path towards sustainability for the Nordic mining industry. J. Clean. Prod. 151, 43−52. https://doi.org/10.1016/j.jclepro.2017.03.047.

Raugei, M., Ulgiati, S., 2009. A novel approach to the problem of geographic allocation of environmental impact in life cycle assessment and material flow analysis. Ecol. Indicat. 9, 1257−1264. https://doi.org/10.1016/j.ecolind.2009.04.001.

Raugei, M., Winfield, P., 2019. Prospective LCA of the production and EoL recycling of a novel type of Li-ion battery for electric vehicles. J. Clean. Prod. 213, 926−932. https://doi.org/10.1016/j.jclepro.2018.12.237.

Renzulli, P.A., Notarnicola, B., Tassielli, G., Arcese, G., Di Capua, R., 2016. Life cycle assessment of steel produced in an Italian integrated steel mill. Sustainability 8. https://doi.org/10.3390/su8080719.

Ruan, S., Unluer, C., 2016. Comparative life cycle assessment of reactive MgO and Portland cement production. J. Clean. Prod. 137, 258−273. https://doi.org/10.1016/j.jclepro.2016.07.071.

Santero, N., Hendry, J., 2016. Harmonization of LCA methodologies for the metal and mining industry. Int. J. Life Cycle Assess. 21, 1543−1553. https://doi.org/10.1007/s11367-015-1022-4.

Stewart, M., Petrie, J., 2006. A process systems approach to life cycle inventories for minerals: South African and Australian case studies. J. Clean. Prod. 14, 1042−1056. https://doi.org/10.1016/j.jclepro.2004.08.008.

Tan, R.B.H., Khoo, H.H., 2005. An LCA study of a primary aluminum supply chain. J. Clean. Prod. 13, 607−618. https://doi.org/10.1016/j.jclepro.2003.12.022.

Teh, S.H., Wiedmann, T., Castel, A., de Burgh, J., 2017. Hybrid life cycle assessment of greenhouse gas emissions from cement, concrete and geopolymer concrete in Australia. J. Clean. Prod. 152, 312−320. https://doi.org/10.1016/j.jclepro.2017.03.122.

Tost, M., Hitch, M., Chandurkar, V., Moser, P., Feiel, S., 2018. The state of environmental sustainability considerations in mining. J. Clean. Prod. 182, 969−977. https://doi.org/10.1016/j.jclepro.2018.02.051.

Van Genderen, E., Wildnauer, M., Santero, N., Sidi, N., 2016. A global life cycle assessment for primary zinc production. Int. J. Life Cycle Assess. 21, 1580−1593. https://doi.org/10.1007/s11367-016-1131-8.

Weidema, B.P., Norris, G.A., 2002. Avoiding co-product allocation in the metals sector. ICMM Int. Work. Life Cycle Assess. Met. https://lca-net.com/files/icmm.pdf.

Weidema, B.P., Bauer, C., Hischier, R., Mutel, C., Nemecek, T., Reinhard, J., Vadenbo, C.O., Wernet, G., 2013. Data Quality Guideline for the Ecoinvent Database Version 3, vol. 3, p. 169.

Weng, Z., Haque, N., Mudd, G.M., Jowitt, S.M., 2016. Assessing the energy requirements and global warming potential of the production of rare earth elements. J. Clean. Prod. 139, 1282−1297. https://doi.org/10.1016/j.jclepro.2016.08.132.

Westfall, L.A., Davourie, J., Ali, M., McGough, D., 2016. Cradle-to-gate life cycle assessment of global manganese alloy production. Int. J. Life Cycle Assess. 21, 1573−1579. https://doi.org/10.1007/s11367-015-0995-3.

Wolf, M.A., Pant, R., Chomkhamsri, K., Sala, S., Pennington, D., 2012. The International Reference Life Cycle Data System (ILCD) Handbook, European Commission. https://doi.org/10.2788/85727. JRC references reports.

World Economic Forum, 2015. Mining & Metals in a Sustainable World 2050, pp. 1–44. REF 250815.

Zaimes, G.G., Hubler, B.J., Wang, S., Khanna, V., 2015. Environmental life cycle perspective on rare earth oxide production. ACS Sustain. Chem. Eng. 3, 237–244. https://doi.org/10.1021/sc500573b.

Zhang, L., Wang, J., Feng, Y., 2018. Life cycle assessment of opencast coal mine production: a case study in Yimin mining area in China. Environ. Sci. Pollut. Res. 25, 8475–8486. https://doi.org/10.1007/s11356-017-1169-6.

Zhao, R., Liu, L., Zhao, L., Deng, S., Li, S., Zhang, Y., Li, H., 2019. Techno-economic analysis of carbon capture from a coal-fired power plant integrating solar-assisted pressure-temperature swing adsorption (PTSA). J. Clean. Prod. 214, 440–451. https://doi.org/10.1016/j.jclepro.2018.12.316.

Chapter 3

# Life cycle Assessment of Ilmenite and Rutile Production in Australia

## Introduction

Ilmenite and rutile are the most commonly found and abundant form of titanium oxide. Ilmenite is weakly magnetic mineral sand, grey-black in colour, solid in form and exists in a triangle crystal structure. On the other hand, rutile is reddish-brown and exists in a tetragonal crystal structure. These heavy mineral sands are excavated and dredged, mostly to commercially produce ilmenite, rutile or other titanium oxide ores. Originally, ilmenite and rutile are titanium oxides which may contain a variable amount of magnesium or manganese, which is often dispersed from its original content (Abzalov, 2016). Ilmenite and rutile, being titanium oxide minerals, are used to produce high-performance metal parts such as artificial human body parts, aircraft engine parts, sporting equipment, synthetic rutile, pigments, etc. These pigments are used for whitening in papers, paints, toothpaste, adhesive, plastic and foods and nanotechnologies. From ilmenite ores, through the Becher process, synthetic rutile is produced, which is slightly yellowish and almost transparent in colour (Pellegrino and Lodhia, 2012; Ranängen and Lindman, 2017; Raugei and Ulgiati, 2009).

Australia has one of the largest resources of ilmenite and rutile forms of titanium oxide, and it is quite expected that Australia is facing the environmental impacts and health effects caused by the titanium ore extraction process (Haque et al., 2014; Jones, 2009; Reichl et al., 2016). The effect of this processing route on human health and ecosystems needs to be identified and minimised to ensure a sustainable, environment-friendly extraction process in the long run. The major environmental issues of concern from titanium oxide mining are pollution of groundwater resources, mineral transport with heavy vehicles, dredging operations in fragile coastal areas and deforestation. Many motives are associated with this question as to why the environmental impacts caused by mineral-sand industries require urgent attention. The first issue is the proper rehabilitation of the abandoned mineral-sand deposits after the

Life Cycle Assessment for Sustainable Mining. https://doi.org/10.1016/B978-0-323-85451-1.00003-2

decommissioning or depletion of the mineral ore bodies because there is a possibility of leaching to groundwater resources from the abandoned mineral deposits. Moreover, elevated radiation hazards are associated with the mineral-sand loading and storage facilities which require advanced control systems. Secondly, there have been significant changes over the past 40 years in the mode of mining industry operations, which further leads to the massive scale of the mining activities increasing the chance of abandoning the mineral-sand deposits (Haque et al., 2014; Northey et al., 2016; Onn and Woodley, 2014). In the next step, the reagents added during the secondary processing of mineral-sand deposits which may include chlorides for titanium oxide processing and increases the radio-nuclide emission from the abandoned mineral sites to local groundwater systems, as mineral-sand extraction involves the extraction of soil up to 30 m depth or more, where the topsoil is removed prior to mining and then filled by the tailings from mining (Jones, 2009; Moran et al., 2014; Norgate and Haque, 2010; Sonter et al., 2014). As a result, mineral-sand mining leads to the loss of trees and plants. All these issues had influenced some previous researchers to investigate the environmental impact assessment caused by the mineral industries with some suggestions for improvement (Durucan et al., 2006; Durucan and Korre, 1998, 2003; Farjana et al., 2018b; Frischknecht et al., 2000; Swensen, 1996). Australia has a great resource for mineral-sand industry, the highest in the entire world, but as compared to the total deposit only a small fraction is currently being extracted. By ensuring an eco-friendly sustainable extraction process, Australia could potentially increase its current production of mineral sands like ilmenite and rutile. To facilitate that process, effective measures should be taken to make the mining extraction processes sustainable (Awuah-Offei and Adekpedjou 2011; Law and Lane, 1991; Mudd, 2009; Navarro and Zhao, 2014; Norgate et al., 2007). Although the mining processes are impactful to the environment in different categories of global warming, greenhouse gas emissions, human health and ecotoxicity, such a critical issue has not been reported anywhere in the open literature that provides a comprehensive analysis of the environmental impacts caused by ilmenite and rutile mining technologies in Australia. This chapter provides the knowledge gap with an effort to promote a sustainable extraction process of ilmenite and rutile in Australia.

## Ilmenite—Rutile Mining and Processing

Hard-rock mining methods are used for the recovery of mineral-sand products in some countries, whereas in Australia, Iluka Resources entail dry mining, which involves the typical extraction of heavy mineral ores from entirely shallow, free-flowing and hollow deposits. Dry mining requires transportation like trucks, loaders, excavators or scrapers to recover ore from the mining plant. This transportation unit then delivers the ore to the wet concentration plant. For large ore bodies with low clay content, the wet method of mining is preferred. To reduce the cost in the wet method, dredging with bucket wheels and suction is done. On the other hand, dry methods involve earth-moving equipment to excavate and transport the mineral sand to a separate feed-preparation section. Pumpers or conveyors differentiate among various dry-mining processes. Another exception involves hydraulic mining using high-pressure water. Dredging is done depending on the variability of the ground condition and the amount of available water. After the rehabilitation studies, the land is cleared, and preparation starts for mining. Topsoil and subsoil are removed, stripped and stockpiled. Transportation, like scrapers or trucks, is utilised to collect and transport the ore from the mining plant. Then the ore is screened to make it free from oversized materials, which may include rocks or debris. These oversized items are then returned to the pit to be transferred to the concentrator plant via conveyor. The wet concentration process produces a higher-grade heavy mineral concentrate. Spiral separators are then used to wash the ore through gravity separation. This process consumes a huge amount of water which is then recycled back to the clean water dam. The concentration process produces a mixture of valuable and nonvaluable heavy sand minerals. From this mixture, ilmenite, rutile, zircon and monazite are then separated through dry processing. Ilmenite is a titanium oxide mineral which is upgraded from 85% to 95% titanium oxide, which in turn comes as rutile. Synthetic rutile production consists of two stages, including pyro-metallurgical processing, while ilmenite is heated in a large rotary kiln. It produces iron oxide impurities within the crystal lattice. In the next stage, irons are removed by oxidation and leaching in hydrometallurgical processing (Jones, 2009; Norgate et al., 2007; Nuss and Eckelman, 2014; Sachs and Investment, 2011).

## Life Cycle Assessment Methodology

The life cycle assessment (LCA) analyses and calculates the environmental effects and impacts caused by the process or product while manufacturing throughout its entire life cycle from the cradle-to-grave, cradle-to-gate, gate-to-grave or whatever. LCA works through identification and quantification of input materials, energy and resources and output products and emissions to the environment like air, water or soil. The LCA methodology in this chapter is based on the International Organization for Standards (ISO 14040) which follows four mandatory stages which are required to be completed: goal and scope definition, life cycle inventory, life cycle impact assessment and interpretation and recommendation of the inventory and results (Durucan et al., 2006; Durucan and Korre, 1998, 2003; Onn and Woodley, 2014).

## Goal and Scope Definition

The goal of this chapter is to analyse the life cycle environmental impact of ilmenite and rutile in Australia. The scope of this chapter is the several impact categories assessed by the ILCD method and 10 impact categories assessed by the Cumulative Energy Demand (CED) method. This chapter also addresses the dependency of the environmental impacts of the mineral-sand extraction processes on the electricity grid mix. The life cycle impact assessment is done using SimaPro software version 8.4. The datasets are collected from the Australian Life cycle Inventory dataset (AusLCI) for different mines in Australia. The analysis methods are chosen as the ILCD method and the CED method. The processes are cradle-to-gate life cycle impact assessment. The functional unit is 1 kg of mineral sands. Figs 4 and 5 illustrate the material flow sheet for ilmenite and rutile processing (Schmidt and Thrane, 2009).

## Life Cycle Inventory Analysis

These datasets are system process datasets where the key components or the emissions are divided under the categories of energy generation related to ilmenite/rutile production, diesel, hard coal, natural gas, heavy fuel and electricity. Fig. 6 graphically represents the system boundary diagram used in this chapter. Table 19 shows the comparative input datasets of the ilmenite and rutile extraction processes. Table 20 provides the outputs of the ilmenite and

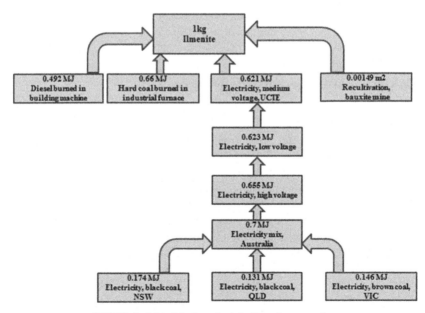

**FIGURE 4** Materials flow sheet for ilmenite processing.

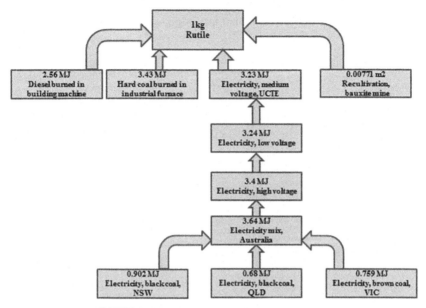

**FIGURE 5** Materials flow sheet for rutile processing.

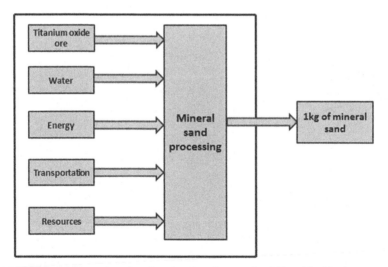

**FIGURE 6** System boundary for mineral-sand processing followed in this chapter.

rutile extraction process where the functional unit for both processes is 1 kg of mineral sand. It also shows the radioactive emissions associated with mining activities such as uranium ions, thorium ions and radium ions (Navarro and Zhao, 2014; Norgate et al., 2007).

**TABLE 19** Input inventory datasets for ilmenite and rutile manufacturing processes.

| Input | From | Ilmenite | Rutile | Unit |
|---|---|---|---|---|
| $TiO_2$ | Nature | 0.56875 | 1 | kg |
| Transformation, from forest | Nature | 0.000594 | 0.00308 | $m^2$ |
| Transformation, from pasture and meadow | Nature | 0.000891 | 0.00462 | $m^2$ |
| Transformation, to the mineral extraction site | Nature | 0.001485 | 0.00771 | $m^2$ |
| Occupation, the mineral extraction site | Nature | 0.01485 | 0.0771 | $m^2$ |
| Water, well, in-ground | Nature | 0.00968 | 0.0502 | $m^3$ |
| Diesel, burned in building machine | Materials/ fuels | 0.286 | 1.4857 | MJ |
| Hard coal, burned in industrial furnace 1–10MW/RER U | Materials/ fuels | 0.66 | 3.4285 | MJ |
| Natural gas, burned in industrial furnace >100 kW/RER U | Materials/ fuels | 0.198 | 1.0285 | MJ |
| Heavy fuel oil, burned in industrial furnace 1 MW, nonmodulating | Materials/ fuels | 0.066 | 0.3428 | MJ |
| Electricity, medium voltage, production UCTE, at grid/UCTE U | Materials/ fuels | 0.16775 | 0.87143 | kWh |
| Mine | Materials/ fuels | 2.20E-10 | 1.14E-9 | p |
| Recultivation | Materials/ fuels | 0.001485 | 0.0077143 | $m^2$ |

## Life Cycle Impact Assessment

The subsections describe the life cycle impact assessment results under various impact analysis criteria based on the ILCD and CED methods. The ILCD method is the International Reference Life Cycle Data System method, which is a standard method based on ISO 14040 and developed by the European Commission, Institute for Environment and Sustainability. This LCIA method considers 14 significant midpoint impact categories. For life cycle analysis in the Australian geographic context, the Australian Indicator method seems to suit best, but it has a major limitation. This method only considers weighting for the climate-change category, but it is a key feature for LCA analysis.

**TABLE 20** Output inventory dataset of ilmenite–rutile manufacturing processes.

| Output | Emissions to | Ilmenite | Rutile | Unit |
|---|---|---|---|---|
| 54% titanium dioxide, at the plant | Product | 1 | 1 | kg |
| Particulates, > 10 μm | Air | 0.000079 | 0.00041 | kg |
| Particulates, > 2.5 μm and < 10 μm | Air | 0.000041 | 0.00021 | kg |
| Particulates, < 2.5 μm | Air | 0.000016 | 0.000085 | kg |
| Uranium-238 | Air | 0.0000091 | 0.000047 | kBq |
| Thorium-232 | Air | 0.000019 | 0.0001 | kBq |
| Radium-226 | Air | 0.0000315 | 0.00016 | kBq |
| Heat, waste | Air | 0.603 | 3.13714 | MJ |
| Suspended solids, unspecified | Water | 0.000025 | 0.00013 | kg |
| Iron, ion | Water | 0.0000012 | 6.57E-6 | kg |

Moreover, the ILCD method is a more robust and complete method and is a combination of CML, ReCiPe, IPCC and EcoTox methods. In the second step, the CED method is used for assessment of the energy (renewable and nonrenewable) consumptions in the mineral-sand processing routes. In the third step, sensitivity analysis is carried out to illustrate the electricity grid mix dependency of mineral-sand extraction processes in Australia. Fig. 7 shows the life cycle impact assessment methodology followed in this chapter based on the ILCD and CED methods (Curran, 2012; Farjana et al., 2018c; Parvez Mahmud et al., 2018; PE International, 2014).

## Comparative Impact Assessment Results using ILCD Method

The detailed comparative impact analysis results from the ilmenite and rutile extraction processes are provided in Table 21. In the climate-change category, climate-change effects are greatest, at 1.53 kg $CO_2$ eq from rutile and 0.295 kg $CO_2$ eq from ilmenite. Ilmenite processing produces 0.000155 kg of PM2.5 eq particulate matter, while rutile processing has impacts like 0.00081 kh PM2.5 eq. In terms of eutrophication, the impact is 0.00469 molc N eq from ilmenite and from rutile, it is 0.024 molc N eq. For human toxicity, the effect is measured on CTUh, and rutile (1.25E-9 CTUh) has effects larger than ilmenite (2.41E-10 CTUh).

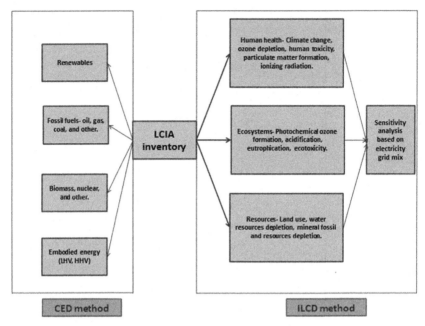

**FIGURE 7** LCA methodology used in this chapter.

Figs 8 and 9 show the impact assessment results using the ILCD method. Fig. 8 graphically represents the normalised results from the ILCD method. Fig. 9 illustrates the single-score results of ilmenite and rutile processing routes, which reveal the weighted scores of all impact categories in a single unit to assess the environmental effects. Among the subsystems of ilmenite mining activities, most of the emissions are due to electricity consumption in mining activities. The second-largest contributor to global-warming emission is hard coal. Similar results are obtained for rutile mining processes. From the comparison between ilmenite and rutile, rutile effects are significantly larger than ilmenite extraction operations. The major impact categories from the life cycle impact assessment of ilmenite—rutile mineral extraction activities are climate change, particulate matter, terrestrial eutrophication, human toxicity noncancer and water resources depletion. Under these categories, electricity generation contributes the most, while the heat generation from diesel contributes to the second most. But the scenario is quite different for the ozone depletion potential of terrestrial ecosystems, while heavy fuel oil contributes the most. Eutrophication is the potential of adding excessive nutrients to water resources or groundwater resources from the mining or manufacturing process. For freshwater eutrophication, the electricity generation of the ilmenite mining process has a severe impact rather than the rutile manufacturing process,

**TABLE 21** Comparative life cycle impact assessment results based on the International Reference Life Cycle Data System method.

| Impact category | Unit | Ilmenite, 54% titanium dioxide | Rutile, 95% titanium dioxide, at the plant |
|---|---|---|---|
| Climate change | kg $CO_2$ eq | 0.2954 | 1.535 |
| Ozone depletion | kg CFC-11 eq | 6.28E-09 | 3.26E-08 |
| Human toxicity, noncancer effects | CTUh | 4.42E-09 | 2.30E-08 |
| Human toxicity, cancer effects | CTUh | 2.41E-10 | 1.25E-09 |
| Particulate matter | kg PM2.5 eq | 0.000155 | 0.00081 |
| Ionising radiation HH | kBq U235 eq | 1.50E-05 | 7.79E-05 |
| Ionising radiation E (interim) | CTUe | 9.37E-13 | 4.87E-12 |
| Photochemical ozone formation | kg NMVOC eq | 0.00123 | 0.00644 |
| Acidification | molc $H^+$ eq | 0.00138 | 0.0072 |
| Terrestrial eutrophication | molc N eq | 0.00469 | 0.0243 |
| Freshwater eutrophication | kg P eq | 1.70E-07 | 8.85E-07 |
| Marine eutrophication | kg N eq | 0.000437 | 0.0022 |
| Freshwater ecotoxicity | CTUe | 0.0175 | 0.091 |
| Land use | kg C deficit | 0.332 | 1.728 |
| Water resource depletion | $m^3$ water eq | 0.00163 | 0.0085 |
| Mineral, fossil and renewable resource depletion | kg Sb eq | 2.05E-16 | 1.06E-15 |

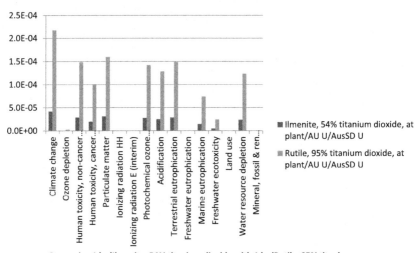

Comparing 1 kg 'Ilmenite, 54% titanium dioxide with 1 kg 'Rutile, 95% titanium dioxide

**FIGURE8** Comparative life cycle impact assessment results based on the ILCD method − normalised results.

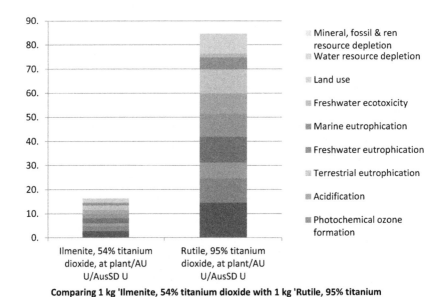

Comparing 1 kg 'Ilmenite, 54% titanium dioxide with 1 kg 'Rutile, 95% titanium dioxide

**FIGURE 9** Comparative life cycle impact assessment results based on the ILCD method − single-score results.

wherein total ilmenite is higher than the rutile mining process. Ecotoxicity is the chemical or biological effect caused by the chemical reactions of the manufacturing or production processes on the ecosystems. The categories considered here are terrestrial ecotoxicity, freshwater ecotoxicity, marine ecotoxicity and marine aquatic ecotoxicity.

In all these categories, the result is consistent: the highest impact is caused by the electricity consumed in the mining and extraction process for both ilmenite and rutile. The second-largest contributor is the heavy fuel oil-based mining process. Overall, rutile has larger environmental impacts than ilmenite. The Human Toxicity Potential (HTP) is a quantitative toxic equivalency potential (TEP) that illustrates the harmful effects of a unit of chemical released. The categories considered here are human toxicity, carcinogenic and noncarcinogenic and the overall human toxicity obtained from the ILCD method. The analysis results are similar to those presented before under various categories, and the results indicate that the electricity consumption for the mineral-sand mining process is the largest contributor for both ilmenite and rutile, with rutile effects larger than ilmenite. The second-largest contributor is the diesel fuel-based mining process which is consistent with some of the findings reported previously.

## Comparative Impact Assessment Results from Cumulative Energy Demand Method

Table 22 represents the LCIA results from the CED, which assessed the breakdown of the fuel used throughout the mineral-sand processing of ilmenite and rutile from the titanium oxide ores. Fig. 10 illustrates the major energy-intensive process. The fuel inputs considered here across the mineral-sand processing systems are fossil fuels, renewables, nuclear, biomass, embodied energy (lower heating values), embodied energy (higher heating values) and other energy sources. Fig. 10 illustrates that the highest environmental threat from mineral processing is due to the usage of fossil fuels, mostly coal. The reason behind this result is the huge amount of electricity consumption in mineral processing. In Australia, the major source of electricity generation is coal, which is found in abundance in many major coal mines located in Australia. The second-largest energy is consumed in other forms of fossil fuels, such as oil and gas. Diesel is consumed in the diesel electric generating unit and also burned in building machines. A comparison between ilmenite and rutile processing is presented in Fig. 10, which shows that rutile is much more energy-intensive mineral sand than ilmenite because the lower ore grade consumes a greater amount of energy extracted from the core of the earth.

**TABLE 22** Comparative life cycle impact assessment results based on Cumulative Energy Demand (CED) method.

| Impact category | Unit | Ilmenite, 54% titanium dioxide | Rutile, 95% titanium dioxide |
|---|---|---|---|
| Renewables | MJ LHV | 0.0789 | 0.41 |
| Fossil fuels – oil | MJ LHV | 0.6828 | 3.547 |
| Fossil fuels – gas | MJ LHV | 0.5621 | 2.92 |
| Fossil fuels – coal | MJ LHV | 2.1113 | 10.968 |
| Fossil fuels – other | MJ LHV | 0 | 0 |
| Biomass | MJ LHV | 0.0006 | 0.00345 |
| Nuclear | MJ LHV | 1.27E-05 | 6.60E-05 |
| Other/unknown | MJ LHV | 0 | 0 |
| Embodied energy LHV | MJ LHV | 3.435 | 17.849 |
| Embodied energy HHV | MJ HHV | 3.52 | 18.286 |

## Sensitivity Analysis of Ilmenite and Rutile Extraction Routes

Tables 23–25 provide the sensitivity analysis results based on the electricity grid mix of different regions. The goal of this analysis is to identify and discuss the effects of electricity on mineral-sand processing, as electricity generation and consumption is the greatest contributor to these effects. Seven different case scenarios are considered including the base case scenario. The electricity grid mixes and their inventories are collected from the EcoInvent database, which is further considered for sensitivity analysis.

In the base case of Australia, the electricity grid mix is considered for UCTE (Union for the Co-ordination of Transmission of Electricity). The other cases are electricity grid mix for CENTRAL (Central European Power Association), CH (Switzerland), DE (Germany), NL (Netherlands), NORDEL (Nordic Countries Power Association) and RER (Europe). Different regional

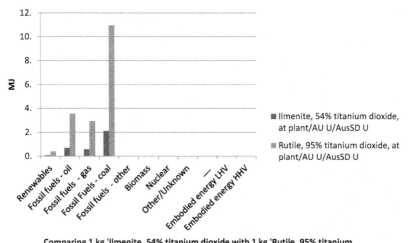

Comparing 1 kg 'Ilmenite, 54% titanium dioxide with 1 kg 'Rutile, 95% titanium dioxide

**FIGURE 10**   Comparative life cycle impact assessment results based on CED method — weighted results.

**TABLE 23** Electricity grid mix details.

| Electricity grid mix | Annual electricity production | Producer |
| --- | --- | --- |
| Australia — UCTE (Union for the Co-ordination of Transmission of Electricity) | 2,577,860 GWh | Germany, France, Spain and Italy |
| CENTRAL (Central European Power Association) | 270,588 GWh | Poland and the Czech Republic |
| NORDEL (Nordic Countries Power Association) | 394,639 GWh | Sweden and Norway |

areas utilise different energy mixes for their electricity generation. The energy mixes vary in terms of renewable energy utilisation for electricity generation, source of fossil fuels utilised in electricity generation, small-scale and large-scale hydroelectric power, etc. Each set of electricity grid mix has specific properties which significantly affects sustainability. From the analysis results presented below in Table 26 for ilmenite, the annual electricity production in the UCTE network is 2,577,860 GWh with Germany, France, Spain and Italy. The annual production in the NORDEL network is 394,639 GWh with Sweden

**TABLE 24** Sensitivity analysis results for ilmenite – based on seven different grid mix scenarios.

| Impact category | Unit | Electricity, at UCTE Total | Electricity, at CENTRAL Total | Electricity, at CH Total | Electricity, at DE Total | Electricity, at NL Total | Electricity, at NORDEL Total | Electricity, at RER Total |
|---|---|---|---|---|---|---|---|---|
| Climate change | kg $CO_2$ eq | 2.95E-01 | 2.74E-01 | 1.30E-01 | 2.35E-01 | 2.46E-01 | 2.96E-01 | 2.08E-01 |
| Ozone depletion | kg CFC-11 eq | 6.28E-09 | 7.25E-09 | 8.36E-09 | 9.21E-09 | 7.02E-09 | 6.28E-09 | 7.85E-09 |
| Human toxicity, noncancer effects | CTUh | 4.42E-09 | 1.07E-08 | 3.19E-09 | 5.90E-09 | 4.78E-09 | 4.42E-09 | 6.02E-09 |
| Human toxicity, cancer effects | CTUh | 2.41E-10 | 1.64E-09 | 1.87E-10 | 4.24E-10 | 2.02E-10 | 2.41E-10 | 4.88E-10 |
| Particulate matter | kg PM2.5 eq | 1.56E-04 | 1.96E-04 | 1.41E-04 | 1.56E-04 | 1.70E-04 | 1.56E-04 | 1.77E-04 |
| Ionising radiation HH | kBq U235 eq | 1.50E-05 | 1.27E-02 | 3.23E-02 | 9.29E-03 | 2.53E-03 | 1.50E-05 | 2.16E-02 |
| Ionising radiation E (interim) | CTUe | 9.37E-13 | 1.14E-07 | 2.54E-07 | 8.48E-08 | 2.27E-08 | 9.37E-13 | 1.95E-07 |
| Photochemical ozone formation | kg NMVOC eq | 0.00123 | 1.10E-03 | 8.18E-04 | 9.13E-04 | 9.72E-04 | 1.24E-03 | 9.70E-04 |

| | | | | | | | | |
|---|---|---|---|---|---|---|---|---|
| Acidification | molc H⁺ eq | 0.00138 | 1.94E-03 | 1.09E-03 | 1.24E-03 | 0.00126 | 0.00138 | 1.49E-03 |
| Terrestrial eutrophication | molc N eq | 4.69E-03 | 3.95E-03 | 2.96E-03 | 3.32E-03 | 3.55E-03 | 4.69E-03 | 0.00349 |
| Freshwater eutrophication | kg P eq | 1.70E-07 | 4.74E-08 | 1.82E-08 | 4.62E-08 | 9.31E-08 | 1.70E-07 | 6.68E-08 |
| Marine eutrophication | kg N eq | 0.000431 | 3.61E-04 | 2.70E-04 | 0.0003 | 0.00032 | 0.000431247 | 0.000319 |
| Freshwater ecotoxicity | CTUe | 0.0175 | 4.10E-02 | 1.53E-02 | 1.90E-02 | 1.68E-02 | 1.75E-02 | 0.025083551 |
| Land use | kg C deficit | 0.3327 | 3.44E-01 | 3.36E-01 | 3.42E-01 | 0.3478 | 0.3327 | 3.42E-01 |
| Water resource depletion | m³ water eq | 0.0016 | 2.44E-03 | 0.00168 | 0.0022 | 0.0019 | 0.0016 | 0.002076 |
| Mineral, fossil and renewable resource depletion | kg Sb eq | 2.05E-16 | 3.11E-16 | 2.54E-16 | 4.61E-15 | 2.22E-15 | 2.05E-16 | 1.68E-15 |

**TABLE 25** Sensitivity analysis results for rutile – based on seven different grid mix scenarios.

| Rutile Impact category | Unit | Electricity, at UCTE Total | Electricity, at CENTRAL Total | Electricity, at CH Total | Electricity, at DE Total | Electricity, at NL Total | Electricity, at NORDEL Total | Electricity, at RER Total |
|---|---|---|---|---|---|---|---|---|
| Climate change | kg $CO_2$ eq | 1.54E+00 | 1.42E+00 | 6.78E-01 | 1.22E+00 | 1.28E+00 | 1.54E+00 | 1.08E+00 |
| Ozone depletion | kg CFC-11 eq | 3.26E-08 | 3.76E-08 | 4.34E-08 | 4.78E-08 | 3.64E-08 | 3.26E-08 | 4.08E-08 |
| Human toxicity, noncancer effects | CTUh | 2.30E-08 | 5.56E-08 | 1.66E-08 | 3.07E-08 | 2.48E-08 | 2.30E-08 | 3.12E-08 |
| Human toxicity, cancer effects | CTUh | 1.25E-09 | 8.51E-09 | 9.71E-10 | 2.20E-09 | 1.05E-09 | 1.25E-09 | 2.53E-09 |
| Particulate matter | kg PM2.5 eq | 8.10E-04 | 1.02E-03 | 7.32E-04 | 8.09E-04 | 8.81E-04 | 0.00081 | 9.21E-04 |
| Ionising radiation HH | kBq U235 eq | 7.79E-05 | 6.59E-02 | 1.68E-01 | 4.83E-02 | 1.32E-02 | 7.79E-05 | 1.12E-01 |
| Ionising radiation E (interim) | CTUe | 4.87E-12 | 5.94E-07 | 1.32E-06 | 4.41E-07 | 1.18E-07 | 4.87E-12 | 1.01E-06 |
| Photochemical ozone formation | kg NMVOC eq | 6.44E-03 | 5.70E-03 | 4.25E-03 | 4.74E-03 | 5.05E-03 | 0.00644 | 5.04E-03 |

| Acidification | molc H+ eq | 0.0072 | 1.01E-02 | 5.65E-03 | 6.42E-03 | 0.0065 | 0.0072 | 7.74E-03 |
|---|---|---|---|---|---|---|---|---|
| Terrestrial eutrophication | molc N eq | 2.44E-02 | 0.0205 | 0.0153 | 0.0172 | 1.85E-02 | 0.0243 | 1.82E-02 |
| Freshwater eutrophication | kg P eq | 8.85E-07 | 2.46E-07 | 9.46E-08 | 2.40E-07 | 4.84E-07 | 8.85E-07 | 3.47E-07 |
| Marine eutrophication | kg N eq | 0.00224 | 1.87E-03 | 1.40E-03 | 1.57E-03 | 1.68E-03 | 0.00224 | 1.66E-03 |
| Freshwater ecotoxicity | CTUe | 9.10E-02 | 2.13E-01 | 7.92E-02 | 9.86E-02 | 8.71E-02 | 0.091001 | 0.130305 |
| Land use | kg C deficit | 1.728695 | 1.79E+00 | 1.74E+00 | 1.77E+00 | 1.806 | 1.728 | 1.777 |
| Water resource depletion | m³ water eq | 8.51E-03 | 0.0126 | 0.0087 | 0.011599 | 1.01E-02 | 0.0085 | 0.01 |
| Mineral, fossil and renewable resource depletion | kg Sb eq | 1.06E-15 | 1.62E-15 | 1.32E-15 | 2.40E-14 | 1.16E-14 | 1.06E-15 | 8.71E-15 |

and Norway. The annual production in the CENTRAL network is 270,588 GWh with Poland and the Czech Republic as the main producers (Frischknecht et al., 2015; Itten et al., 2014). The annual production in the RER network corresponds to the production in the ENTSOE network excluding the Baltic States. There are some electricity losses due to transmission, which is dependent on the population density in the different countries and on the infrastructure of technology. The statistics of global electricity production show that, in Australia, around 92.2% of the total electricity production is based on fossil fuels and around 5% of production is from hydropower. In Switzerland, fossil fuel-based electricity generation is only 2.45%, while most of the electricity generation is from hydropower (around 60%). In Germany, fossil fuel only generates 60% of the total electricity production, while the rest comes from nuclear power (around 23%) and hydropower (around 5%). The Netherlands produce electricity based on fossil fuels (around 85%). After briefly analysing the electricity mixes from different technologies and fuel sources for different countries, it is evident that Australia is one of the leading countries utilising fossil fuels for electricity generation (Frischknecht et al., 2015; Itten et al., 2014).

The analysis results presented in Tables 24 and 25 and Fig. 11 clarify the fact that the electricity grid mix based on UCTE is one of the greatest impactful grid mix scenarios. Table 27 shows the sensitivity analysis results of ilmenite processing based on seven different grid mix scenario. The results show that sensitivity varies depending on the grid mix scenario and impact category. NORDEL electricity mix is harmful to the environment over the other grid mixes in the categories of climate change, acidification, eutrophication, ecotoxicity and land use. Similar results come from UCTE network, which means for Australia, UCTE grid mix sensitivity results are almost

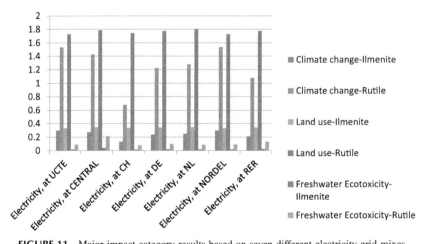

**FIGURE 11** Major impact category results based on seven different electricity grid mixes.

similar like NORDEL. CENTRAL electricity mixes are harmful to human toxicity (cancer effects) and particulate matter. Switzerland electricity mix is impactful on ionising radiation. In Europe (RER), it has significant impacts on human toxicity. Electricity mix of Germany is impactful for ozone depletion and mineral resources depletion. Among the seven electricity grid mix scenarios considered for ilmenite processing routes, the electricity grid mix used in the Netherlands is the most sustainable. This is due to the lowest consumption of fossil fuels. The highest impact is caused by the electricity grid mix in the NORDEL network and UCTE network due to more than 80% being fossil fuel based. Among all the impact categories assessed by the ILCD method, the most sensitive categories due to electricity generations are climate change, land use and freshwater ecotoxicity (Farjana et al., 2018c,e; Frischknecht et al., 2015; Itten et al., 2014; Paraskevas et al., 2016; Santero and Hendry, 2016; Tost et al., 2018; Tuusjärvi et al., 2014; Vintró et al., 2014).

Table 25 shows the sensitivity analysis results of rutile processing based on seven different grid mix scenario. The results show that sensitivity varies depending on the grid mix scenario and impact category. NORDEL electricity mix is harmful to the environment over the other grid mixes in the categories of climate change, acidification, eutrophication, ecotoxicity and photochemical ozone formation. UCTE grid mix sensitivity results are almost similar, like NORDEL one which is impactful for climate change, eutrophication and acidification. CENTRAL electricity mixes are harmful for human toxicity (cancer effects and noncancer effects) and water resources depletion. Switzerland electricity mix is impactful on ionising radiation. In Europe (RER), it has significant impacts on freshwater ecotoxicity. Among the seven electricity grid mix scenarios considered for rutile processing routes, the electricity grid mix used in Germany is the most sustainable. Among the 14 major impact categories assessed by the ILCD method for rutile processing, the most sensitive categories due to electricity generations are climate change, land use and freshwater ecotoxicity. As rutile processing is more energy-intensive due to ore grades, it consumes more electricity and has a greater effect on sustainability (Farjana et al., 2018c,e; Frischknecht et al., 2015; Itten et al., 2014; Paraskevas et al., 2016; Santero and Hendry, 2016; Tost et al., 2018; Tuusjärvi et al., 2014; Vintró et al., 2014).

Fig. 11 illustrates the comparison among the sensitivity analysis results of different categories (climate change, land use, freshwater ecotoxicity) for the seven different electricity grid mix scenario. Both ilmenite and rutile are considered and compared here. The analysis results indicate the level of uncertainty among the impact values. In terms of climate change, both ilmenite processing and rutile processing are more sustainable in Switzerland over other case scenarios. Land use is another affected category from these mineral processing routes, while NORDEL shows the highest level of sustainability. In the case of freshwater ecotoxicity, UCTE network shows a higher level of sustainability. The reasons behind these varied analysis results are the consumption of fossil fuels used for the electricity generation. Because in ilmenite and rutile processing, minerals consume a huge amount of electricity to be produced, this has enormous effects on sustainability. Integration of renewable energy resources to the mineral processing systems can contribute to making the mineral miner countries more sustainable (Farjana et al., 2018a,c,d).

## Conclusion

In summary, this chapter addresses the environmental impacts caused by the ilmenite and rutile extraction processes through detailed midpoint indicator-based LCA, material flow analysis and sensitivity analysis based on electricity grid mix of different global regions. The significance of this chapter is to assess the sustainability issues of the mineral-sand industry in Australia, as Australia has one of the largest reserves and production of ilmenite and rutile. The life cycle impact analysis results of ilmenite and rutile processing routes and the corresponding sensitivity analysis based on the electricity grid mix show that a significant environmental impact is caused by the rutile mining process over the ilmenite mining process. The reason behind this is that ilmenite is produced from 54% titanium oxide with a higher ore grade where rutile is produced from 95% titanium oxide with lower ore grade. Extracting the low-grade rutile is more energy-intensive than extracting the higher-grade ilmenite in addition to the fact that rutile is extracted from deep mines. The analysis results indicate that the major contributor for both the ilmenite and rutile mining process is electricity consumption because a significant amount of electricity is consumed during the mining and milling processes of mineral-sand industries. An exception is found for the rutile production processes in the eutrophication category, where diesel energy-based process is the greatest contributor. The hard coal-based production phase is in the second position for global warming potential, acidification potential and eutrophication. The diesel-based production process is the second-largest contributor for ozone formation and human toxicity. For ecotoxicity, heavy fuel oil effects are significantly greater than the diesel-oil and hard coal-based production phases. Overall, the results found from this chapter show that a significant environmental hazard is generated from the mineral-sand industries in Australia, which will continue to increase because of the depletion of higher-grade minerals, which will lead to deeper mining of lower-grade mineral sands. Finally, the results presented and discussed in this chapter elucidate some valuable insights into the impacts generated from the mineral-sand industries in Australia and will surely help to minimise some of the impacts by taking appropriate measures.

# References

Abzalov, M., 2016. Mineral sands. Mod. Approaches Solid Earth Sci. 12, 427−433. https://doi.org/10.1007/978-3-319-39264-6_35.

Awuah-Offei, K., Adekpedjou, A., 2011. Application of life cycle assessment in the mining industry. Int. J. Life Cycle Assess. 16, 82−89. https://doi.org/10.1007/s11367-010-0246-6.

Curran, M.A., 2012. Life Cycle Assessment Handbook. https://doi.org/10.1002/9781118528372.

Durucan, S., Korre, A., 1998. Life cycle assessment of mining projects for waste minimisation and long term control of rehabilitated sites. Environ. Toxicol. 1001−1008.

Durucan, S., Korre, A., 2003. Mining Life Cycle Modelling for Environmental Control and Waste Minimisation, pp. 3−10.

Durucan, S., Korre, A., Munoz-Melendez, G., 2006. Mining life cycle modelling: a cradle-to-gate approach to environmental management in the minerals industry. J. Clean. Prod. 14, 1057−1070. https://doi.org/10.1016/j.jclepro.2004.12.021.

Farjana, S.H., Huda, N., Mahmud, M.A.P., 2018a. Environmental Impact Assessment of European Non-ferro Mining Industries through Life-Cycle Assessment Environmental Impact Assessment of European Non-ferro Mining Industries through Life-Cycle Assessment 0−7.

Farjana, S.H., Huda, N., Mahmud, M.A.P., 2018b. Life-cycle environmental impact assessment of mineral industries. In: IOP Conference Series: Materials Science and Engineering. https://doi.org/10.1088/1757-899X/351/1/012016.

Farjana, S.H., Huda, N., Mahmud, M.A.P., Saidur, R., 2018c. Solar process heat in industrial systems − a global review. Renew. Sustain. Energy Rev. 82 https://doi.org/10.1016/j.rser.2017.08.065.

Farjana, S.H., Huda, N., Mahmud, M.A.P., Saidur, R., 2018d. Solar industrial process heating systems in operation − current SHIP plants and future prospects in Australia. Renew. Sustain. Energy Rev. 91 https://doi.org/10.1016/j.rser.2018.03.105.

Farjana, S.H., Huda, N., Parvez Mahmud, M.A., 2018e. Environmental impact assessment of european non-ferro mining industries through life-cycle assessment. In: IOP Conference Series: Earth and Environmental Science, vol. 154, p. 012019. https://doi.org/10.1088/1755-1315/154/1/012019.

Frischknecht, R., Braunschweig, A., Hofstetter, P., Suter, P., 2000. Human health damages due to ionising radiation in life cycle impact assessment. Environ. Impact Assess. Rev. 20, 159−189. https://doi.org/10.1016/S0195-9255(99)00042-6.

Frischknecht, R., Itten, R., Sinha, P., Wild-Scholten, M. de, Zhang, J., Fthenakis, V., Kim, H.C., Raugei, M., Stucki, M., 2015. Life Cycle Inventories and Life Cycle Assessments of Photovoltaic Systems. PVPS Task 12. International Energy Agency. Report T12-04:2015.

Haque, N., Hughes, A., Lim, S., Vernon, C., 2014. Rare earth elements: overview of mining, mineralogy, uses, sustainability and environmental impact. Resources 3, 614−635. https://doi.org/10.3390/resources3040614.

PE International, 2014. Harmonization of LCA Methodologies for Metals: A Whitepaper Providing Guidance for Conducting LCAs for Metals and Metal Products.

Itten, R., Frischknecht, R., Stucki, M., Scherrer, P., Psi, I., 2014. Life Cycle Inventories of Electricity Mixes and Grid. Paul Scherrer Inst, pp. 1−229. https://doi.org/10.13140/RG.2.2.10220.87682.

Jones, G., 2009. Mineral Sands: An Overview of the Industry. Iluka, pp. 38−41 unpublished.

Law, E., Lane, D.A., 1991. Boston College International and Comparative Law Review Pollution Caused by Waste from the Titanium Dioxide Industry: Directive 89/428. C. Int'l Comp. L. Rev 14428.

Moran, C.J., Lodhia, S., Kunz, N.C., Huisingh, D., 2014. Sustainability in mining, minerals and energy: new processes, pathways and human interactions for a cautiously optimistic future. J. Clean. Prod. 84, 1—15. https://doi.org/10.1016/j.jclepro.2014.09.016.

Mudd, G.M., 2009. The Sustainability of Mining in Australia: Key Production Trends and Their Environmental Implications for the Future. ISBN: 978-0-9803199-4-1.

Navarro, J., Zhao, F., 2014. Life-cycle assessment of the production of rare-earth elements for energy applications: a review. Front. Energy Res. 2, 1—17. https://doi.org/10.3389/fenrg.2014.00045.

Norgate, T., Haque, N., 2010. Energy and greenhouse gas impacts of mining and mineral processing operations. J. Clean. Prod. 18, 266—274. https://doi.org/10.1016/j.jclepro.2009.09.020.

Norgate, T.E., Jahanshahi, S., Rankin, W.J., 2007. Assessing the environmental impact of metal production processes. J. Clean. Prod. 15, 838—848. https://doi.org/10.1016/j.jclepro.2006.06.018.

Northey, S.A., Mudd, G.M., Saarivuori, E., Wessman-Jääskeläinen, H., Haque, N., 2016. Water footprinting and mining: where are the limitations and opportunities? J. Clean. Prod. 135, 1098—1116. https://doi.org/10.1016/j.jclepro.2016.07.024.

Nuss, P., Eckelman, M.J., 2014. Life cycle assessment of metals: a scientific synthesis. PLoS One 9, 1—12. https://doi.org/10.1371/journal.pone.0101298.

Onn, A.H., Woodley, A., 2014. A discourse analysis on how the sustainability agenda is defined within the mining industry. J. Clean. Prod. 84, 116—127. https://doi.org/10.1016/j.jclepro.2014.03.086.

Paraskevas, D., Kellens, K., Van De Voorde, A., Dewulf, W., Duflou, J.R., 2016. Environmental impact analysis of primary aluminium production at country level. Procedia CIRP 40, 209—213. https://doi.org/10.1016/j.procir.2016.01.104.

Parvez Mahmud, M.A., Huda, N., Farjana, S.H., Asadnia, M., Lang, C., 2018. Recent advances in nanogenerator-driven self-powered implantable biomedical devices. Adv. Energy Mater. 8, 1—25. https://doi.org/10.1002/aenm.201701210.

Pellegrino, C., Lodhia, S., 2012. Climate change accounting and the Australian mining industry: exploring the links between corporate disclosure and the generation of legitimacy. J. Clean. Prod. 36, 68—82. https://doi.org/10.1016/j.jclepro.2012.02.022.

Ranängen, H., Lindman, Å., 2017. A path towards sustainability for the Nordic mining industry. J. Clean. Prod. 151, 43—52. https://doi.org/10.1016/j.jclepro.2017.03.047.

Raugei, M., Ulgiati, S., 2009. A novel approach to the problem of geographic allocation of environmental impact in life cycle assessment and material flow analysis. Ecol. Indicat. 9, 1257—1264. https://doi.org/10.1016/j.ecolind.2009.04.001.

Reichl, C., Schatz, M., Zsak, G., 2016. World mining data 2016. Fed. Minist. Sci. Res. Econ. Austria 31, 1—255.

Sachs, G., Investment, A., 2011. Mineral Deposits — Corporate Snapshot, pp. 1—25.

Santero, N., Hendry, J., 2016. Harmonization of LCA methodologies for the metal and mining industry. Int. J. Life Cycle Assess. 21, 1543—1553. https://doi.org/10.1007/s11367-015-1022-4.

Schmidt, J.H., Thrane, M., 2009. Life Cycle Assessment of Aluminium Production in New Alcoa Smelter in Greenland University 2009.

Sonter, L.J., Moran, C.J., Barrett, D.J., Soares-Filho, B.S., 2014. Processes of land use change in mining regions. J. Clean. Prod. 84, 494—501. https://doi.org/10.1016/j.jclepro.2014.03.084.

Swensen, G., 1996. The Management of Radiation Hazards from the Mining of Mineral Sands in Western Australia, vol. 17, pp. 1—16.

Tost, M., Hitch, M., Chandurkar, V., Moser, P., Feiel, S., 2018. The state of environmental sustainability considerations in mining. J. Clean. Prod. 182, 969–977. https://doi.org/10.1016/j.jclepro.2018.02.051.

Tuusjärvi, M., Mäenpää, I., Vuori, S., Eilu, P., Kihlman, S., Koskela, S., 2014. Metal mining industry in Finland-development scenarios to 2030. J. Clean. Prod. 84, 271–280. https://doi.org/10.1016/j.jclepro.2014.03.038.

Vintró, C., Sanmiquel, L., Freijo, M., 2014. Environmental sustainability in the mining sector: evidence from Catalan companies. J. Clean. Prod. 84, 155–163. https://doi.org/10.1016/j.jclepro.2013.12.069.

Chapter 4

# Comparative Life Cycle Assessment of Uranium Extraction Processes

## Introduction

Uranium is the heaviest metal naturally found in the earth's crust which is used as a nuclear fuel throughout the world. It is a radioactive, ductile and dense metal which forms uranium oxide when it comes in contact with air. It is used widely for producing nuclear energy through atomic chain reactions, making shields in bullets or missiles and producing fertilisers. Among the various natural isotopes of uranium available on the earth, the uranium-238 isotope constitutes more than 98%, while the rest is the uranium-235 isotope. The typical form of uranium found in the mines is uranium oxide in a 3 g per tonne concentration. To be an economically viable extraction site, the concentration must be 300 mg/kg (Mudd and Diesendorf, 2008). As a power generation source, nuclear fuel has a range of economic benefits accompanied by environmental benefits. First of all, it offers a competitive fuel price considering the capital cost of the plant for power generation despite a costly licencing system. In comparison with coal or natural gas, nuclear fuel only comprises 15% of the operating costs. Secondly, nuclear fuel is a low-carbon power generation source. However, there were some atomic accidents with detrimental health effects: the Chernobyl and Fukushima nuclear accidents, for example, that caused some devastating effects on the environment and human health. Uranium oxide is a source of nuclear power with emissions lower than fossil fuel-based energy generation systems (Ashley et al., 2015; Koltun, 2014; Lenzen, 2008; Mudd and Diesendorf, 2010; Mutchek et al., 2016; Norgate et al., 2014; Poinssot et al., 2014; Warner and Heath, 2012). More than 400 atomic reactors have been operating throughout the world for electricity generation since 2012, which necessitates driving the uranium mining system towards a more sustainable and eco-friendly extraction system (Norgate et al., 2014; Parker et al., 2016; Vintró et al., 2014). Due to the recent surge in the usage of uranium around the world, environmental impacts of nuclear fuel mining technologies should be assessed thoroughly to ensure sustainability

Life Cycle Assessment for Sustainable Mining. https://doi.org/10.1016/B978-0-323-85451-1.00004-4

during the electricity generation from nuclear sources. To ensure the sustainability of uranium extraction technologies from the manufacturer's perspective, it is indispensable to perform an environmental effects assessment throughout the entire life cycle of the uranium manufacturing process. The life cycle environmental emissions of nuclear power generation are dependent on uranium mining technologies. The source of electricity used for the uranium extraction process could, accordingly, contribute to environment-friendly nuclear fuel generation (Awuah-Offei and Adekpedjou, 2011; Hudson et al., 1999; Schmidt and Thrane, 2009). Life cycle assessment (LCA) is a powerful environmental impact assessment tool that is used for calculating the impacts of manufacturing technologies on various categories of human health, ecosystems and resources of the environment (Curran, 2012). This chapter will address the life cycle environmental impact assessment for three different mining process of uranium extraction: open-pit, underground and in-situ leaching.

## Uranium Mining Process

The mining technologies that are used in extracting uranium are broadly classified in three ways: in-situ leaching/solution mining, open-pit and underground mining. Underground mining techniques were the most common one during the early years of uranium extraction. In later years, this contribution decreased noticeably with the invention of the in-situ leaching process. The World Nuclear Association states that 48% of the total uranium production comes from in-situ leaching mines. Table 26 provides a summary of current uranium extraction methods.

**TABLE 26** Brief description of uranium extraction methods (World Nuclear Association, 2014).

| Methods of uranium ore mining | Description | World production (%) by the World Nuclear Association |
|---|---|---|
| Underground mining | Vertical shafts or tunnels are used for drilling depending on the depth of the deposit. | 47 |
| Open-cut mining | A large amount of overburden must be excavated to reach the uranium ore. | |
| In-situ leaching | Vertical boreholes are used to introduce a leaching solution to further extract uranium from shallow deposits. | 48 |
| Recovery from the mining of other materials | To recover uranium from other material mines, detailed analysis about the depth of overburden, depth of ore body and inclination of the ore body is required to choose which method would be economical. | 5 |

In the open-pit mining method, mining involves drilling and blasting, excavation, haulage and sorting. The milling operation constitutes of crushing and grinding, leaching, washing, clarifying and filtering, solvent extraction, precipitation and wet-cake drying and calcination. The scope of this process covers the production of uranium yellowcake during the operation of an open-pit uranium mine, from resource extraction to the transportation from the mill to the conversion facility. After mining the finely ground uranium, it goes through a conventional chemical-based extraction process which requires leaching, concentration and purification. The leaching technique is dependent on the uranium ore grade. It might be dynamic lixiviation (average recovery of 95% of uranium) or static lixiviation (typical improvement of 60%−80% of uranium). The in-situ leaching method is used in some mining sites and involves an entirely different process from the previous ones (ELaw, 2014). The next step is to concentrate the uranium through ion exchange or extraction. Chemical precipitation is done later after concentrating the uranium oxide which contains impurities. This product is known as yellowcake. Finally, yellowcake is heated to get rid of ammonia and to make 97% pure uranium oxide. Then the yellowcake is transferred to purification or other processing facilities through ship, truck or train depending on the mine facilities. Purification is necessary for further enrichment of uranium to modify it into solid uranium oxide which is used as the nuclear fuel in nuclear power generation units. After the uranium enrichment process, depleted uranium waste is produced, which can be further processed to be used as an alternative nuclear fuel. In addition to this, large quantities of waste rock are produced that contain low-grade uranium (Lunt et al., 2007). This waste rock, in turn, is converted into radionuclides, which are higher in uranium concentration and become radioactive waste (Aden et al., 2010; Mudd, 2006). The advantage of in-situ mining is that it allows nondisturbed surface extraction of uranium as the process involves injecting reagents through a small diameter hole sink for mobilising the minerals. Then the extracted solution is processed to recover the metals. Another significant advantage is that tailing storage is not required. Additionally, it never consumes massive amounts of water due to continuous recycling through the injected wells. Due to the benefits of in-situ leaching uranium mining methods, it might be expected that the environmental burdens would be significantly reduced; however, the reinjected fluid can potentially spread to other groundwater resources which could be dangerous (Mudd, 2000; Taylor et al., 2004).

## Sustainability Challenges of Uranium Mining

Uranium mines should satisfy the environmental risk management criteria, under the categories of human health, environment, land uses and water resources, to be commissioned as operating mines to align with sustainability policies. The hazardous effects and emissions from these uranium mines should be assessed throughout their entire life cycle (Taylor et al., 2004). Over the last decades, tailings and waste management in uranium mills have been a focus, and regulations are also changing for environmental hazards and radiological risks. To reveal the regulatory changes, an examination of history is required. During the 1950s, wastes from mines were discharged to lowland areas constituting rivers and water resources. These had a significant effect in wet seasons as the water quality decreased and erosion occurred in wet regions. In arid regions, uranium mills' radioactive and mineral wastes were managed through water dams. During the 1970s, regulations were introduced to transfer the tailings back into the mined pit as soon as possible after uranium extraction, by using these ground dams. From the 1980s in-situ leach mining methods were tested on a massive scale, depending on the leaching type: for example, acid leaching in Honeymoon mine and alkaline leaching at the Manyingee mine. The acid leaching method was developed through the Beverly mine in 2001. The waste rock and uranium tailings should have been a focus of attention as a source of radioactive waste. Nonetheless, these areas received less attention despite being a significant source of radon emission. In addition to producing radioactive wastes, these large amounts of wastes pose a health threat towards the community living close to uranium mines and mills. The process of making yellowcake involves highly toxic substances. These radioactive tailings could contaminate the groundwater resources through rain and erosion. Contamination could also occur through insufficient covering of wastes and tailings. From the heaps of materials, wind could blow radioactive particles away from the mines. Wind erosion also occurs due to the inadequate coverage of tailings. The amount of water required for uranium mining is huge, and it is returned to groundwater resources like lakes and rivers. This mixture of fresh groundwater and water used in radioactive processing has direct detrimental health effects, also damaging the ecosystems which it encounters. This reduces freshwater availability and at the same time causes significant risks to the animals and plants due to radioactivity. If uranium mines are abandoned, these mines will be flooded with water, which also implies contamination of freshwater. Leaking from all these sources may cause harm to respiratory systems in the human body, lung cancer, leukaemia, stomach cancer and congenital disabilities (Mudd, 2000, Mudd and Diesendorf, 2008; Northey et al., 2013; World Nuclear Association, 2014). Attention should also be paid to the use of radioactive by-product metals for several nonmining purposes, such as the manufacture of kitchen utensils (in which the health risks associated with the use of these radioactive metals remain) (Mudd, 2006). At earth's surface, on all continents, there are some levels of natural radiation in the soil, water, air, flora and fauna. Uranium-238, uranium-235 and their daughter products are present in all rocks and soil. These natural radionuclides are also naturally present in the lower atmosphere (dust from the soil) along with radon-222 which permanently emanates from the rocks and soil. They are also present in surface and underground waters in contact with this soil and rocks as well as in the crops, flora and fauna and therefore in the food chain. The radiation emitted by these radionuclides is called natural background radiation. When uranium ore is buried underground at a depth of few tens or even a few hundreds of metres, the radiation levels at the surface of the soil remain low and usually have the same order of magnitude as natural radiation levels. Some areas of a limited extent (a few square metres) can be found where the ore reaches the ground surface. Otherwise, the protection offered by the soil is usually sufficient to reduce the risks for people living in the area. Indeed, alpha and low-energy beta particles are stopped by a thin layer of soil (much less than 1 cm). Even penetrating gamma radiation does not cross a layer of soil only a few metres deep. Most of the radon gas remains trapped inside the soil because of its short half-life (3.8 days), and many of the gas atoms will disintegrate inside the soil during their migration before reaching the biosphere. When uranium is mined, uranium ores with high uranium content are brought to the surface. A typical ore with a uranium concentration of 0.2% has a uranium-238 activity of about 25,000 Bq/kg. The total activity calculated including all the uranium-238 daughter products and the uranium-235 decay chain, therefore, exceeds 360,000 Bq/kg, while the mean activity of the earth's crust is below 2000 Bq/kg. Such material should be managed with a great deal of caution due to the risks of exposure to ionising radiation (Norgate et al., 2014; Northey et al., 2013, 2016; Sonter et al., 2014).

## Materials and Method

LCA is a systematic tool for analysing and calculating the environmental effects and impacts caused by the manufacturing of a product, process or activity throughout its entire life cycle from the cradle-to-grave, cradle-to-gate and gate-to-grave. LCA works through identification and quantification of input materials, energy and resources and output products and emissions to the environment such as air, water or soil. The LCA methodology in this chapter is based on the International Organization for Standards (ISO) 14040, that follows four mandatory stages which are required to be completed: goal and scope definition, life cycle inventory, life cycle impact assessment and interpretation and recommendation of the inventory and results (Durucan et al., 2006; Onn and Woodley, 2014). SimaPro software is used for analysis and comparison for different uranium extraction routes is presented in details. The datasets are considered from EcoInvent databases, AusLCI and several literature resources.

## Goal and Scope Definition

The goal of this chapter is to assess the life cycle environmental impacts of uranium extraction routes: open-pit mining, underground mining and in-situ leaching. This chapter also compares its LCA results to identify the most sustainable uranium extraction process. The LCA is done under a cradle-to-gate perspective for those three processes, which considers from raw materials acquisition to the packaging and delivery of the final product (Drielsma et al., 2016; Santero and Hendry, 2016). Figs 12—14 sequentially illustrate the energy and heat consumption in different stages of the uranium extraction process for open-pit mining, underground mining and in-situ leaching, respectively. The functional unit assessed in this chapter is 1 kg of uranium, which is the representative of the unit amount of reference flow (Ranängen and Lindman, 2017; Raugei and Ulgiati, 2009).

Lenzen et al. thoroughly evaluated the life cycle energy balance of nuclear power generation in Australia. They also compared the energy balance of the nuclear fuel cycle with the energy balance of other renewable and nonrenewable power generation sources (Lenzen, 2008). Parker et al. (2016) worked on life cycle greenhouse gas emission calculations and assessments of uranium mining and milling in Canada. Haque et al. analysed greenhouse gas emissions from the in-situ leaching-based uranium extraction process and then compared

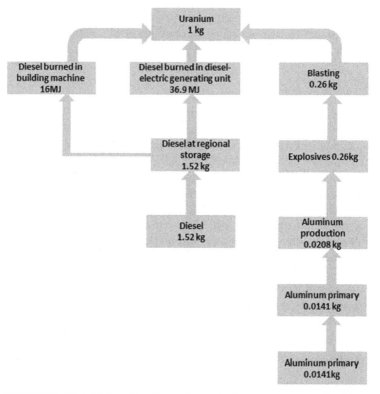

**FIGURE 12**  Material flow sheet for uranium extraction process — open-pit mining.

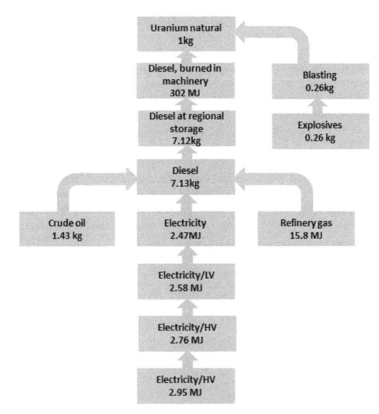

**FIGURE 13**  Material flow sheet for the uranium extraction process — underground mining.

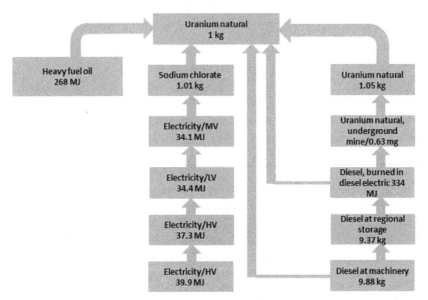

**FIGURE 14**  Material flow sheet for the uranium extraction process — in-situ leaching mining.

this with the conventional uranium mining process (Haque and Norgate, 2014; Norgate et al., 2014). Moran et al. (2014) outlined the sustainability of mining and energy with the focus on future mining challenges, sustainability framework and mining management aspects. Pellegrino et al. worked on climate-change accounting in Australia and explored the critical factors of the Australian mining industry regarding climate change (Pellegrino and Lodhia, 2012). Schneider et al. (2013a,b) proposed a new research framework for energy, land and water use in the uranium mining and milling industries. Tost et al. (2018) reviewed the state-of-the-art environmental sustainability considerations in mining, mostly non-LCA. Tuusjärvi et al. (2014) outlined the Finnish mining industries' developmental scenario and considered the scenario analysis concerning sustainability. Mudd et al. carried out comprehensive research in several papers based on sustainability assessment of uranium mining, in-situ leaching mining, radiation hazards and emissions from uranium mining (Mudd, 2000, 2008, 2002; Mudd and Diesendorf, 2010, 2008; Northey et al., 2016). Haque et al. assessed the life cycle impacts of in-situ leaching uranium mining and the effects of uranium ore grade in their works (Haque and Norgate, 2014; Norgate et al., 2014). Table 27 summarises

**TABLE 27** Literature survey on sustainability research on uranium mining technologies.

| Literature | Analysis method | Description |
|---|---|---|
| Johnston and Needham (1998) | Non-LCA | Environmental protection near the area of Ranger mine, Australia, provides a summary of all biological monitoring methods. |
| Mudd (2000) | Non-LCA | A detailed and comprehensive discussion on the change of ionising radiation and radionuclide releases from former and operational projects on uranium mining in Australia. |
| CSIRO report Taylor et al. (2004) | Non-LCA | Research and review based on the fact that in-situ acid leaching would contaminate groundwater. |
| Mudd (2008) | Non-LCA | Researches on energy, water consumption and carbon emission in respect of uranium production. |
| Mudd and Diesendorf (2008) | Non-LCA | Detailed compilation and analysis of radon releases in Australia. |
| Mudd and Diesendorf (2010) | Non-LCA | Comprehensive insight into uranium mining and its economic resources. |
| Haque and Norgate (2014) | LCA | Life cycle assessment of in-situ leaching uranium mining methods collected the global data and integrated Australian electricity mix. |

the literature survey of sustainability research on uranium mining technologies. So far, life cycle impact assessment of uranium mines, comparing the relative benefits of open-pit mining, underground mining and in-situ leaching has not been reported in the open literature. Furthermore, there is no detailed research on an environmental impact comparison of uranium production methods. The current work thus serves as a vital resource to provide such invaluable information.

## Life Cycle Inventory Analysis

This stage identifies the materials, resources and energy inputs, output products, wastes and emissions of uranium production and quantifies their respective amounts per unit process included in the system boundary for LCA. For carrying out the LCA, the second stage is to collect the inventory data. The inventories for the uranium extraction process vary among different companies, countries and regions. This variation can eventually affect impact assessment results. The main difference among the inventories here is their raw materials, energy inputs, fossil fuels and electricity consumption. The input flows can be defined as the raw materials, energy and resources and are given for 1 kg of uranium yellowcake production. Fig. 15 shows the system boundary for the LCA of uranium extraction routes. The system boundary encloses uranium ore, raw materials, fuels, energy, electricity, inorganic and organic chemicals and transportation.

On the other hand, uranium yellowcake (1 kg) and waste emission are considered as the outputs (Durucan and Korre, 2003; Heard, 2017). Table 28

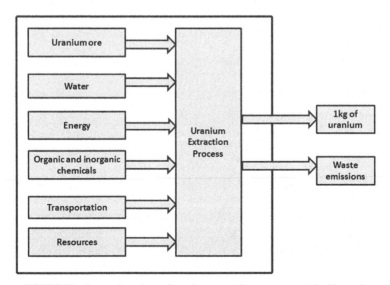

**FIGURE 15**   System boundary of uranium extraction process used in this work.

**TABLE 28** Life cycle inventory datasets — inputs for uranium extraction process.

| Inputs | From | Open-pit mining method | Underground mining method | In-situ leaching mining method | Unit |
|---|---|---|---|---|---|
| Transformation, from unknown | Nature | 0.182 | 0.005 | 0.018 | $m^2$ |
| Transformation, to the mineral extraction site | Nature | 0.182 | 0.005 | 0.018 | $m^2$ |
| Transformation, from an industrial area | Nature | 0.182 | 0.005 | | $m^2$ |
| Transformation, to unknown | Nature | 0.182 | 0.005 | | $m^2$ |
| Occupation, the mineral extraction site | Nature | 4 | 0.1 | 1.96 | $m^2a$ |
| Uranium, in-ground | Nature | 1.05 | 1.05 | 1.05 | kg |
| Water, unspecified natural origin | Nature | 6 | 0.1 | 1 | $m^3$ |
| Diesel, burned in building machine | Materials/fuel | 16 | | | MJ |
| Diesel, burned in diesel-electric generating set | Materials/fuel | 36.5 | 300 | 176 | MJ |
| Blasting | Materials/fuel | 0.26 | 0.26 | | kg |
| Transport, lorry >16t | Materials/fuel | 7.23 | 0.23 | | tkm |
| Transport, freight, rail | Materials/fuel | 1.37 | 1.37 | | tkm |
| Uranium | Materials/fuel | 6.17E-8 | 0.000000278 | 0.000000135 | p |
| Heavy fuel oil, burned in the electric furnace | Materials/fuel | | | 264 | MJ |
| Ammonia, liquid | Materials/fuel | | | 0.9 | kg |

**TABLE 28** Life cycle inventory datasets — inputs for uranium extraction process.—cont'd

| Inputs | From | Open-pit mining method | Underground mining method | In-situ leaching mining method | Unit |
|---|---|---|---|---|---|
| Ammonium sulphate | Materials/ fuel | | | 0.106 | kg |
| Chemicals, inorganic | Materials/ fuel | | | 0.26 | kg |
| Chemicals, organic | Materials/ fuel | | | 0.315 | kg |
| Ethylenediamine | Materials/ fuel | | | 0.012 | kg |
| Soda, powder | Materials/ fuel | | | 2.5 | kg |
| Sodium chlorate, powder | Materials/ fuel | | | 1 | kg |
| Sodium chloride, brine solution | Materials/ fuel | | | 2.5 | kg |
| Sodium hydroxide, 50% in the H$_2$O production mix | Materials/ fuel | | | 0.026 | kg |
| Sulphric acid, liquid | Materials/ fuel | | | 35 | kg |
| Transport, lorry >16t, fleet average | Materials/ fuel | | | 6.3 | tkm |
| Transport, freight, rail | Materials/ fuel | | | 32 | tkm |

provides a life cycle inventory datasheet for input flows (materials and their respective quantities). Table 29 presents life cycle inventory datasheet for output flows (elements and their respective amounts).

## Life Cycle Impact Analysis

The life cycle impact assessment in the analysis phase is based on different methods such as International Reference Life Cycle Data System (ILCD),

**TABLE 29** Life cycle inventory datasets — outputs for uranium extraction process.

| Outputs | To | Quantity in open-pit mining method | Quantity in underground mining method | Quantity in in-situ leaching mining method | Unit |
|---|---|---|---|---|---|
| Uranium, natural | Product | 1 | 1 | 1 | kg |
| Aldehydes, unspecified | Air | | | 0.00088 | kg |
| Ammonia | Air | | | 0.0017 | kg |
| Beryllium | Air | | 0.00000098 | | kg |
| Cadmium | Air | | 0.000002 | | kg |
| Lead | Air | | 0.00002 | 2 | kBq |
| Polonium-210 | Air | | | W | kBq |
| Nitrogen oxides | Air | | | 0.017 | kg |
| NMVOC, nonmethane volatile organic compounds, unspecified origin | Air | | | 0.11 | kg |
| Particulates, > 10 μm | Air | 0.056 | 0.033 | 0.22 | kg |
| Particulates, > 2.5 μm, and <10 μm | Air | | | 0.00088 | kg |
| Sulphur dioxide | Air | | | 0.00023 | kg |
| Zinc | Air | | 0.00002 | | kg |
| Radon-222 | Air | 13,000 | 1,000,000 | 150,000 | kBq |
| Radium-226 | Air | | 13 | 1 | kBq |
| Uranium alpha | Air | 0.094 | 23 | | kBq |
| Thorium-230 | Air | | | 1 | kBq |
| Uranium-234 | Air | | | 2.9 | kBq |
| Uranium-235 | Air | | | 0.14 | kBq |
| Uranium-238 | Air | | | 2.9 | kBq |

**TABLE 29** Life cycle inventory datasets – outputs for uranium extraction process.—cont'd

| Outputs | To | Quantity in open-pit mining method | Quantity in underground mining method | Quantity in in-situ leaching mining method | Unit |
|---|---|---|---|---|---|
| Aluminium | Water | 0.0031 | 0.023 | 0.35 | kg |
| Ammonium, ion | Water | 0.0085 | 0.065 | 0.072 | kg |
| Arsenic, ion | Water | 0.000093 | 0.000068 | 0.000081 | kg |
| Barium | Water | 0.0019 | 0.0014 | 0.00011 | kg |
| Beryllium | Water | | | 0.000014 | kg |
| Cadmium, ion | Water | 0.000093 | 0.000068 | | kg |
| Calcium, ion | Water | | | 0.54 | kg |
| Chloride | Water | 0.86 | 0.6 | 0.52 | kg |
| Chromium, ion | Water | | | 0.00096 | kg |
| Copper, ion | Water | | | 0.0002 | kg |
| Cyanide | Water | | | 0.00000088 | kg |
| Fluoride | Water | | | 0.00066 | kg |
| Hydrocarbons, unspecified | Water | | | 0.0035 | kg |
| Iron, ion | Water | 0.034 | 0.11 | 0.1 | kg |
| Lead | Water | 0.018 | 0.013 | 0.0015 | kg |
| Magnesium | Water | 0.11 | 0.082 | 0.01 | kg |
| Manganese | Water | 0.069 | 0.00075 | 0.015 | kg |
| Molybdenum | Water | 0.0016 | 0.018 | 0.0019 | kg |
| Nickel, ion | Water | | | 0.0001 | kg |
| Nitrate | Water | 0.0021 | 0.041 | 0.0087 | kg |
| Phosphate | Water | | | 0.00022 | kg |
| Selenium | Water | 0.00019 | 0.00041 | 0.0016 | kg |
| Silver, ion | Water | | | 0.00000088 | kg |
| Sodium, ion | Water | | 4.7 | 0.04 | kg |
| Sulphate | Water | 48 | 3.4 | 1.6 | kg |

*Continued*

**TABLE 29** Life cycle inventory datasets — outputs for uranium extraction process.—cont'd

| Outputs | To | Quantity in open-pit mining method | Quantity in underground mining method | Quantity in in-situ leaching mining method | Unit |
|---|---|---|---|---|---|
| Sulphide | Water | | | 0.000044 | kg |
| Suspended solids, unspecified | Water | 2.1 | 0.15 | | kg |
| Titanium, ion | Water | | | 0.0012 | kg |
| Vanadium, ion | Water | 0.0065 | 0.0048 | 0.000018 | kg |
| Zinc, ion | Water | 0.0012 | 0.00089 | 0.00052 | kg |
| Carbonate | Water | | | 0.036 | kg |
| Radium-226 | Water | 5000 | 1600 | 1 | kBq |
| Thorium-230 | Water | 460 | 300 | 150 | kBq |
| Uranium alpha | Water | 220 | 310 | 4.85 | kBq |
| Uranium-234 | Water | | | 0.3 | kBq |
| Uranium-238 | Water | | | 4.85 | kBq |
| Tailings, uranium milling | Waste | | | 0.25 | M3 |

TRACI, CML, ReCiPe, etc. This phase assesses the impacts of the process/ system/activity on human health, ecosystem, water, land and economy. Furthermore, there are optional additions for normalisation, grouping or weighting of the indicator results and data quality analysis techniques. In this chapter, first of all, the life cycle impact analysis (LCIA) is done using ILCD method, which is a standard method based on ISO 14040 and developed by the European Commission Institute for Environment and Sustainability. This LCIA method considers 16 significant midpoint impact categories, including ionising radiation. Ionising radiation is a significant impact category during uranium extraction that provides valuable information that can be assessed during the LCIA using the ILCD method (Adiansyah et al., 2017; Sonter et al., 2014). In the second stage, the Cumulative Energy Demand (CED) method is also utilised to assess the environmental effects based on energy consumption in different stages/materials. The forms of energy considered here are renewables, fossil fuels, biogas, nuclear and others. In the third stage, sensitivity analysis is carried out to identify the critical manufacturing material/

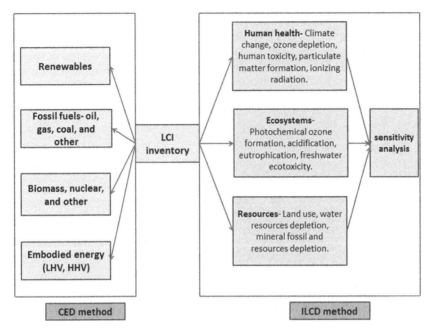

**FIGURE 16**   LCIA methodology of this research.

process steps responsible for environmental emissions to enhance the efficiency towards sustainability (Awuah-Offei and Adekpedjou, 2011; PE International, 2014). Fig. 16 graphically represents the methodology of LCA considered in this chapter.

## Open-pit Mining

Table 30 describes the quantified results of LCIA for open-pit mining. Among the 14 major impact categories presented here, ionising radiation has the most significant impact of 178.55 kBq U235 eq. The second most impactful category is human toxicity (noncancer effects) at 2.56E-06 CTUh. Thirdly, water resources depletion impact is higher than the other impact categories, which is 0.9744 m$^3$ water eq. From the other significant categories, the climate-change effect is 6.27 kg CO$_2$.

Fig. 17 illustrates the LCA results assessed through the ILCD method. From Fig. 17, in open-pit mining, based on the LCIA conducted using the ILCD method, diesel burned in types of machinery is the source of significant impacts for climate change, particulate matter, marine eutrophication and water resource depletion. Diesel burned in the diesel-electric generating unit is responsible for climate change, ozone depletion and ionising radiation (interim). Uranium ore has effects on human toxicity (cancer), human toxicity (noncancer) and ionising radiation. Blasting is most impactful on sustainability

**TABLE 30** Life cycle impact analysis results for open-pit mining using the ILCD method.

| Label | Unit | Total | Uranium | Diesel, machine | Diesel-electric generating set | Blasting | Transport, lorry >16t | Transport, freight |
|---|---|---|---|---|---|---|---|---|
| Climate change | kg CO$_2$ eq | 6.273 | 0 | 22.19 | 50.1034 | 12.34 | 14.8 | 0.556 |
| Ozone depletion | kg CFC-11 eq | 7.02E-07 | 0 | 24.9154 | 55.4327 | 2.779 | 16.4298 | 0.44 |
| Human toxicity, noncancer effects | CTUh | 2.56E-06 | 96.34 | 0.5067 | 1.715 | 1.084 | 0.338 | 0.012 |
| Human toxicity, cancer effects | CTUh | 1.68E-08 | 36.55 | 1.22 | 2.68 | 58.88 | 0.61 | 0.038 |
| Particulate matter | kg PM2.5 eq | 0.00857 | 0 | 21.26 | 35.59 | 40.287 | 2.671 | 0.183 |
| Ionising radiation HH | kBq U235 eq | 178.55 | 99.99 | 1.58E-05 | 3.51E-05 | 6.81E-05 | 1.04E-05 | 2.8E-07 |
| Ionising radiation (interim) | CTUe | 1.15E-09 | 0 | 2.123 | 4.647 | 91.817 | 1.374 | 0.0371 |
| Photochemical ozone formation | kgNMVOCeq | 0.189 | 0 | 10.438 | 29.93 | 54.752 | 4.62 | 0.24 |
| Acidification | molc H$^+$ eq | 0.18 | 0 | 8.4 | 25.31 | 62.62 | 3.45 | 0.19 |
| Terrestrial eutrophication | molc N eq | 0.907 | 0 | 8.0263 | 24.25 | 64.41 | 3.1 | 0.19 |

| | | | | | | | | |
|---|---|---|---|---|---|---|---|---|
| Freshwater eutrophication | kg P eq | 4.38E-06 | 0 | 6.57 | 10.17 | 80.148 | 2.812 | 0.283 |
| Marine eutrophication | kg N eq | 0.0722 | 9.807 | 9.193 | 27.82 | 49.39 | 3.56 | 0.2208 |
| Freshwater ecotoxicity | CTUe | 12.065 | 95.467 | 0.583 | 1.30 | 2.212 | 0.423 | 0.0116 |
| Land use | kg C deficit | 78.424 | 99.714 | 0.0008 | 0.001 | 0.157 | 0.0005 | 0.125 |
| Water resource depletion | m³ water eq | 0.9744 | 99.74 | 0.047 | 0.104 | 0.067 | 0.0308 | 0.0011 |
| Mineral, fossil and resource depletion | kg Sb eq | 4.92E-14 | 0 | 10.625 | 23.15 | 59.19 | 6.84 | 0.18 |

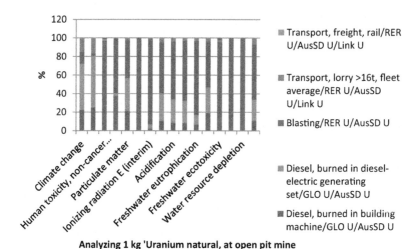

Analyzing 1 kg 'Uranium natural, at open pit mine

**FIGURE 17** Comparative life cycle assessment results — open-pit mining.

that harms human toxicity (cancer effects), particulate matter, photochemical ozone formation, acidification, terrestrial eutrophication, freshwater eutrophication, marine eutrophication, land use, water resource depletion and mineral fossil and resources depletion.

## Underground Mining

Table 31 describes the quantified results of LCIA for underground mining. Among the 14 major impact categories presented here, ionising radiation has the most significant impact at 1150.27 kBq U235 eq. The second most impactful category is human toxicity (noncancer effects) at 4.08E-06 CTUh followed by water resources depletion impacts at 0.0252 $m^3$ water eq. Another notable impact category is the climate-change effect which is 26.67 kg $CO_2$ per kg of uranium production.

Fig. 18 illustrates the LCA results assessed through the ILCD method. As the figure shows, diesel burned in the diesel-electric generating unit is responsible for climate change, ozone depletion, particulate matter, photochemical ozone formation, marine eutrophication and ionising radiation (interim). Uranium ore effects on human toxicity (cancer), human toxicity (noncancer), marine eutrophication, freshwater ecotoxicity, land use and ionising radiation.

## In-situ Leaching

Table 32 summarises the quantified results of LCIA for in-situ leaching mining process. Among the 14 major impact categories presented in the table, ionising radiation has the most significant impact at 989.87 kBq U235 eq. The second most impactful category is human toxicity (noncancer effects) at 5.71E-06

**TABLE 31** Life cycle impact analysis results for underground mining using the ILCD method.

| Impact category | Unit | Total | Uranium | Diesel | Blasting | Transport, lorry >16t | Transport, freight |
|---|---|---|---|---|---|---|---|
| Climate change | kg $CO_2$ eq | 26.67 | 0 | 25.83 | 0.77 | 0.029 | 0.034 |
| Ozone depletion | kg CFC-11 eq | 3.22E-06 | 0 | 3.20E-06 | 1.95E-08 | 3.67E-09 | 3.11E-09 |
| Human toxicity, noncancer effects | CTUh | 4.08E-06 | 3.69E-06 | 3.61E-07 | 2.78E-08 | 2.76E-10 | 3.07E-10 |
| Human toxicity, cancer effects | CTUh | 1.90E-08 | 5.37E-09 | 3.71E-09 | 9.92E-09 | 3.28E-12 | 6.43E-12 |
| Particulate matter | kg PM2.5 eq | 0.028 | 0 | 0.025 | 0.003 | 7.28E-06 | 1.57E-05 |
| Ionising radiation HH | kBq U235 eq | 1150.2 | 1150.27 | 0.0005 | 0.0001 | 5.90E-07 | 5.01E-07 |
| Ionising radiation E (interim) | CTUe | 1.50E-09 | 0 | 4.40E-10 | 1.06E-09 | 5.03E-13 | 4.27E-13 |
| Photochemical ozone formation | kg NMVOC eq | 0.569 | 0 | 0.4655 | 0.103 | 0.0002 | 0.0004 |
| Acidification | molc H⁺ eq | 0.487 | 0 | 0.3745 | 0.112 | 0.00019 | 0.0003 |
| Terrestrial eutrophication | molc N eq | 2.39 | 0 | 1.8 | 0.58 | 0.0008 | 0.0017 |
| Freshwater eutrophication | kg P eq | 7.19E-06 | 0 | 3.66E-06 | 3.51E-06 | 3.92E-09 | 1.24E-08 |
| Marine eutrophication | kg N eq | 0.26 | 0.0598 | 0.165 | 0.035 | 8.18E-05 | 0.0001 |
| Freshwater ecotoxicity | CTUe | 17.2 | 15.66 | 1.29 | 0.26 | 0.0016 | 0.0014 |
| Land use | kg C deficit | 2.23 | 2 | 0.011 | 0.123 | 1.36E-05 | 0.0983 |
| Water resource depletion | m³ water eq | 0.025 | 0.0162 | 0.008 | 0.0006 | 9.54E-06 | 1.11E-05 |
| Mineral, fossil and renewable resource depletion | kg Sb eq | 1.23E-13 | 0 | 9.37E-14 | 2.91E-14 | 1.07E-16 | 9.10E-17 |

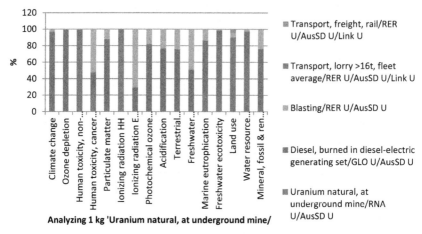

**FIGURE 18** Comparative life cycle assessment results — underground mining.

CTUh followed by human toxicity (cancer effects) impacts up to 6.27E-07 CTUh. From the other significant categories, the climate-change effect is 73.89 kg $CO_2$ and water resources depletion impact is 0.9318 $m^3$ water eq.

Fig. 19 illustrates the LCA results assessed through the ILCD method for in-situ leaching mining technologies for uranium extraction. As the figure shows, uranium natural is responsible for human toxicity (noncancer effects), ionising radiation, photochemical ozone formation, terrestrial eutrophication, marine eutrophication and water resources depletion. Soda powder used in the plant is responsible for climate change, ozone depletion, particulate matter and freshwater ecotoxicity. Ethylenediamine used in the plant is responsible for climate change, ozone depletion, terrestrial eutrophication and marine eutrophication. Inorganic chemicals affect ionising radiation and mineral, fossil and renewable energy resources depletion.

## Results Interpretation and Discussion

This section evaluates and interprets the results from the inventory analysis and impact assessment phase to compare among the production stages and estimate the impacts per production phase, impacts on ecosystems, environment and human health. The LCA can propose a systematic evaluation of the environmental consequences, estimate the amount of emission to the water, air and soil environment in each cycle of the product or process, measure ecological effects of substances and contrast the health and environmental effects of the same product or process in order to select the optimum method (Farjana et al., 2018c; Mahmud et al., 2018).

**TABLE 32** Life cycle impact analysis results for in-situ leaching mining using the ILCD method.

| Label | Unit | Total | Uranium | Diesel burned in the diesel-electric generating set | Heavy fuel oil | Ammonia, liquid, at the regional storehouse | Ammonium sulphate, as N, at the regional storehouse | Chemicals inorganic, at the plant | Chemicals organic, at the plant | Ethylenediamine, at the plant | Soda, powder, at the plant | Sodium chlorate, powder | Sodium chloride, brine solution | Sodium hydroxide, 50% in $H_2O$ |
|---|---|---|---|---|---|---|---|---|---|---|---|---|---|---|
| Climate change | kg $CO_2$ eq | 73.895 | 0 | 20.50 | 32.091 | 2.373 | 0.336 | 0.758 | 0.855 | 0.1229 | 1.50 | 7.84 | 0.31 | 0.066 |
| Ozone depletion | kg CFC-11 eq | 7.90E-06 | 0 | 23.73 | 37.063 | 1.538 | 0.088 | 0.652 | 0.333 | 0.338 | 0.13 | 0.28 | 0.06 | 0.005 |
| Human toxicity, noncancer effects | CTUh | 5.71E-06 | 5.88 | 3.711 | 16.8 | 0.732 | 0.049 | 0.510 | 0.225 | 0.2862 | 0.87 | 1.19 | 0.05 | 0.226 |
| Human toxicity, cancer effects | CTUh | 6.29E-07 | 0.08 | 0.34 | 4.52 | 0.2 | 0.01 | 1.62 | 0.56 | 0.0648 | 0.45 | 49.9 | 0.01 | 0.022 |
| Particulate matter | kgPM2.5eq | 0.11 | 0.22 | 12.9 | 35.6 | 1.3 | 0.13 | 0.26 | 0.20 | 0.021 | 0.83 | 0.49 | 0.07 | 0.00 |
| Ionising radiation HH | kBq U235 eq | 989.8 | 19.1 | 3.05E-05 | 4.65E-05 | 1.6E-05 | 3.76E-07 | 0.0005 | 6.72E-06 | 1.42E-07 | 1.08E-05 | 1.4E-06 | 8.05E-08 | 6.54E-09 |
| Ionising radiation E (interim) | CTUe | 4.88E-08 | 0 | 0.52 | 1.17 | 2.6 | 0.002 | 91.1 | 1.06 | 0.0217 | 0.02 | 0.204 | 0.002 | 0.0001 |
| Photochemical ozone formation | kg NMVOC eq | 1.002 | 12.6 | 27.2 | 5.55 | 0.26 | 0.05 | 0.15 | 0.19 | 0.0237 | 0.32 | 1.4 | 0.05 | 0.01 |
| Acidification | molc H+ eq | 1.468 | 1.22 | 14.9 | 13.3 | 0.36 | 0.04 | 0.28 | 0.13 | 0.0175 | 0.83 | 0.7 | 0.04 | 0.006 |

Continued

**TABLE 32** Life cycle impact analysis results for in-situ leaching mining using the ILCD method.—cont'd

| Label | Unit | Total | Uranium | Diesel burned in the diesel-electric generating set | Heavy fuel oil | Ammonia, liquid, at the regional storehouse | Ammonium sulphate, as N, at the regional storehouse | Chemicals inorganic, at the plant | Chemicals organic, at the plant | Ethylenediamine, at the plant | Soda, powder, at the plant | Sodium chlorate, powder | Sodium chloride, brine solution | Sodium hydroxide, 50% in $H_2O$ |
|---|---|---|---|---|---|---|---|---|---|---|---|---|---|---|
| Terrestrial eutrophication | molc N eq | 3.50 | 2.71 | 30.2 | 4.42 | 0.22 | 0.04 | 0.20 | 0.14 | 0.0305 | 0.97 | 1.7 | 0.05 | 0.01 |
| Freshwater eutrophication | kg P eq | 0.001 | 6.17 | 0.18 | 0.28 | 0.34 | 0.002 | 5.64 | 0.30 | 0.0183 | 4.08 | 0.6 | 0.01 | 0.01 |
| Marine eutrophication | kg N eq | 0.41 | 15.7 | 23.5 | 3.45 | 0.19 | 0.03 | 0.12 | 0.10 | 0.055 | 0.36 | 1.4 | 0.04 | 0.01 |
| Freshwater ecotoxicity | CTUe | 93.5 | 13.9 | 0.80 | 36.5 | 1.52 | 0.07 | 0.19 | 0.25 | 0.0328 | 0.09 | 3.4 | 0.06 | 0.001 |
| Land use | kg C deficit | 167 | 39.4 | 0.004 | 0.002 | 0.02 | 0.01 | 0.09 | 0.004 | 0.001 | 0.11 | | 0.17 | |
| Water resource depletion | m³ water eq | 0.93 | 17.3 | 0.52 | 0.82 | 0.05 | 0.002 | 0.27 | 0.43 | 0.0552 | 1.32 | 2.83 | 0.17 | 0.03 |
| Mineral, fossil and resource depletion | kg Sb eq | 1.24E-11 | 0 | 0.44 | 1.09 | 0.05 | 0.002 | 97.3 | 0.007 | 0.0005 | 0.01 | 0.08 | 0.0019 | 4.7E-05 |

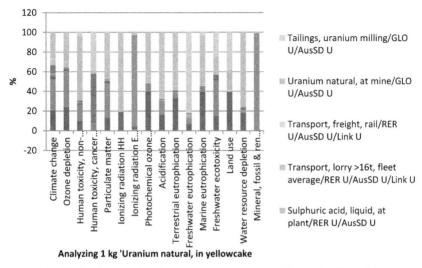

**Analyzing 1 kg 'Uranium natural, in yellowcake**

**FIGURE 19** Comparative life cycle assessment results of in-situ leaching mining.

## Comparative Assessment of Life cycle Assessment Results

Fig. 20 presents a comparative assessment of the life cycle environmental impacts from open-pit mining, underground mining and the in-situ leaching uranium extraction processes. Among the analysis results from 14 impact categories, in-situ leaching uranium extraction has a greater impact than

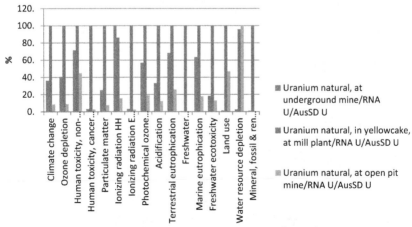

**Comparing 1 kg 'Uranium natural, at underground mine/RNA U/AusSD U', 1 kg 'Uranium natural, in yellowcake, at mill plant/RNA U/AusSD U' and 1 kg 'Uranium natural, at open pit mine/RNA U/AusSD U';**
**Method: ILCD 2011 Midpoint+ V1.10 / EC-JRC Global, equal we**

**FIGURE 20** Graphical representation of ILCD method analysis results.

open-pit mining and underground mining throughout 13 impact categories, which excludes ionising radiation. For ionising radiation, the underground mining method is the most detrimental process. Underground mining holds the second position overall, while the most sustainable uranium extraction process identified is open-pit mining. The reason behind this result is the radon-222 emission from underground mining, which is 1,000,000 kBq, while from open-pit mining radon-222 emission is 13,000 kBq and from in-situ leaching, it is 150,000 kBq. Similarly, the radium-226 discharge is 23 kBq from underground mining, while from open-pit mining radium-226 emission is one kBq and from in-situ leaching, it is 0.094 kBq.

## Comparative Assessment of Results using the Cumulative Energy Demand Method

Fig. 21 graphically represents the LCIA results from the CED method which assessed the breakdown of the fuel used throughout the uranium extraction processes analysed for open-pit mining, underground mining and in-situ leaching. The fuel inputs considered here across the system are fossil fuels, renewables, nuclear, biomass, embodied energy (lower heating values), embodied energy (higher heating values) and other energy sources. In-situ leaching uranium extraction is the most energy-intensive process and consumes the highest amount of fossil fuels, renewables, biomass, nuclear energy and embodied energy. Open-pit mining uranium absorbed the highest amount of atomic energy and embodied the energy, and underground mining consumes the least amount of energy throughout the uranium extraction process.

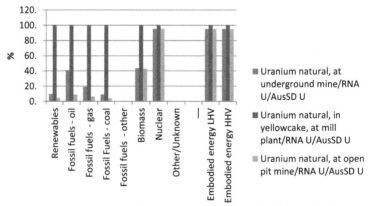

Comparing 1 kg 'Uranium natural, at underground mine/RNA U/AusSD U', 1 kg 'Uranium natural, in yellowcake, at mill plant/RNA U/AusSD U' and 1 kg 'Uranium natural, at open pit mine/RNA U/AusSD U';...

**FIGURE 21** Graphical representation of CED method analysis results.

## Sensitivity Analysis

For the sensitivity analysis performed here, three different case scenarios are chosen, including the base case scenario. The details are outlined below:

**Case 1.**
base case with the required amount of diesel burned in the diesel-electric generating unit.

**Case 2.**
the base case is increased by 50%.

**Case 3.**
the base case is reduced by 50%.

From the sensitivity analysis results shown in Table 33, increasing the energy consumption will significantly increase the negative impacts on climate change, ozone depletion and particulate matter. The sensitivity remains unchanged in ionising radiation, human toxicity (cancer and noncancer effects) and most of the categories of resources. Thermal energy or electrical energy generation sources could be replaced by renewable energy resources, nanopower generation sources or by increasing energy efficiency measures reduce the fossil fuel consumption to reduce the harmful and impactful emissions to the environment regarding land, water, soil and air (Farjana et al., 2018a,b,c; Mahmud et al., 2018; Parvez Mahmud et al., 2018).

## Conclusion

Life cycle inventory data analysis and impact assessment were carried out in this chapter using the commercial software SimaPro version 8.5 using the ILCD and CED methods. The LCIA results show that the underground mining method has greater radioactivity effects than open-pit mining and in-situ leaching mining due to higher quantities of radon-222 and radium-226 emissions. In other major impact categories such as human health (cancer and noncancer), ecotoxicity, resources and climate change, the in-situ leaching method has severe environmental impacts. Due to the higher amount of energy consumption and inorganic and organic chemical consumption (illustrated by the LCIA assessed through the CED method), in-situ leaching mining is more detrimental to the environment than others, though reducing the energy consumption can reduce the environmental effects to a moderate extent, which is further clarified by the sensitivity analysis reported in this chapter. The limitation of this study is that the energy generation in in-situ leaching mining is from heavy fuel oil and the diesel-electric generating unit, which makes the higher impact analysis results. The operating mines where natural gas is used

**TABLE 33** Sensitivity analysis results for uranium extraction processes based on three different case scenarios.

| Label | Open-pit mining | | | Underground mining | | | In-situ leaching | | |
|---|---|---|---|---|---|---|---|---|---|
| | Case 1 | Case 2 | Case 3 | Case 1 | Case 2 | Case 3 | Case 1 | Case 2 | Case 3 |
| Climate change | 90.01 | 100 | 80.02 | 67.37 | 100 | 34.74 | 90.69 | 100 | 81.39 |
| Ozone depletion | 88.92 | 100 | 77.84 | 66.84 | 100 | 33.69 | 89.39 | 100 | 78.78 |
| Human toxicity, noncancer effects | 99.74 | 100 | 99.49 | 95.75 | 100 | 91.51 | 98.17 | 100 | 96.35 |
| Human toxicity, cancer effects | 99.39 | 100 | 98.78 | 91.10 | 100 | 82.21 | 99.82 | 100 | 99.65 |
| Particulate matter | 90.38 | 100 | 80.77 | 69.48 | 100 | 38.97 | 93.92 | 100 | 87.841 |
| Ionising radiation HH | 100 | 100 | 100 | 100 | 100 | 100 | 100 | 100 | 100. |
| Ionising radiation E (interim) | 98.94 | 100 | 97.89 | 87.19 | 100 | 74.39 | 99.73 | 100 | 99.47 |
| Photochemical ozone formation | 95.03 | 100 | 90.07 | 70.99 | 100 | 41.99 | 88.00 | 100 | 76.016 |
| Acidification | 95.96 | 100 | 91.92 | 72.26 | 100 | 44.52 | 93.03 | 100 | 86.078 |
| Terrestrial eutrophication | 96.14 | 100 | 92.28 | 72.59 | 100 | 45.19 | 86.86 | 100 | 73.72 |
| Freshwater eutrophication | 96.81 | 100 | 93.63 | 79.69 | 100 | 59.39 | 99.90 | 100 | 99.817 |
| Marine eutrophication | 95.60 | 100 | 91.21 | 75.95 | 100 | 51.91 | 89.45 | 100 | 78.901 |
| Freshwater ecotoxicity | 99.70 | 100 | 99.41 | 96.38 | 100 | 92.77 | 99.59 | 100 | 99.194 |
| Land use | 99.99 | 100 | 99.99 | 99.73 | 100 | 99.47 | 99.99 | 100 | 99.995 |
| Water resource depletion | 99.97 | 100 | 99.95 | 85.81 | 100 | 71.62 | 99.73 | 100 | 99.476 |
| Mineral, fossil and renewable resource depletion | 94.95 | 100 | 89.91 | 72.42 | 100 | 44.84 | 99.77 | 100 | 99.558 |

as the energy generation source for in-situ leaching, all the impacts will be significantly lower and can be considered as sustainable. Usage of natural gas was not considered in this chapter due to the lack of availability of sufficient data source. If natural gas were considered here, the results would have significant variation, which can ensure greater sustainability during the in-situ leaching mining process. The outcomes of this research provide valuable insights into identifying the most sustainable uranium extraction process among all the significant impact categories studied such as climate change, ionising radiation, human toxicity (cancer) and human toxicity (noncancer) effects. In future, research should be carried out to model and validate appropriate datasets for process and mine-specific uranium extraction methods.

# References

Aden, N., Marty, A., Muller, M., 2010. Comparative Life-Cycle Assessment of Non-fossil Electricity Generation Technologies: China 2030 Scenario Analysis.

Adiansyah, J.S., Haque, N., Rosano, M., Biswas, W., 2017. Application of a life cycle assessment to compare environmental performance in coal mine tailings management. J. Environ. Manag. 199, 181−191. https://doi.org/10.1016/j.jenvman.2017.05.050.

Ashley, S.F., Fenner, R.A., Nuttall, W.J., Parks, G.T., 2015. Life-cycle impacts from novel thorium-uranium-fuelled nuclear energy systems. Energy Convers. Manag. 101, 136−150. https://doi.org/10.1016/j.enconman.2015.04.041.

Awuah-Offei, K., Adekpedjou, A., 2011. Application of life cycle assessment in the mining industry. Int. J. Life Cycle Assess. 16, 82−89. https://doi.org/10.1007/s11367-010-0246-6.

Curran, M.A., 2012. Life Cycle Assessment Handbook. https://doi.org/10.1002/9781118528372.

Drielsma, J.A., Russell-Vaccari, A.J., Drnek, T., Brady, T., Weihed, P., Mistry, M., Simbor, L.P., 2016. Mineral resources in life cycle impact assessment—defining the path forward. Int. J. Life Cycle Assess. 21, 85−105. https://doi.org/10.1007/s11367-015-0991-7.

Durucan, S., Korre, A., 2003. Mining Life Cycle Modelling for Environmental Control and Waste Minimisation 3−10.

Durucan, S., Korre, A., Munoz-Melendez, G., 2006. Mining life cycle modelling: a cradle-to-gate approach to environmental management in the minerals industry. J. Clean. Prod. 14, 1057−1070. https://doi.org/10.1016/j.jclepro.2004.12.021.

ELaw, 2014. Overview of mining and its impacts 1. Guideb. Eval. Min. Proj. EIAs 3−18.

Farjana, S.H., Huda, N., Mahmud, M.A.P., Saidur, R., 2018a. Solar industrial process heating systems in operation − current SHIP plants and future prospects in Australia. Renew. Sustain. Energy Rev. 91 https://doi.org/10.1016/j.rser.2018.03.105.

Farjana, S.H., Huda, N., Mahmud, M.A.P., Saidur, R., 2018b. Solar process heat in industrial systems − a global review. Renew. Sustain. Energy Rev. 82 https://doi.org/10.1016/j.rser.2017.08.065.

Farjana, S.H., Huda, N., Mahmud, M.A.P., 2018c. Environmental Impact Assessment of European Non-ferro Mining Industries through Life-Cycle Assessment, 0−7.

Haque, N., Norgate, T., 2014. The greenhouse gas footprint of in-situ leaching of uranium, gold and copper in Australia. J. Clean. Prod. 84, 382−390. https://doi.org/10.1016/j.jclepro.2013.09.033.

Heard, B., 2017. Environmental Impacts of Uranium Mining in Australia History, Progress and Current Practice Environmental Impacts of Uranium Mining in Australia.

Hudson, T.L., Fox, F.D., Plumlee, G.S., 1999. Metal Mining and the Environment.

Johnston, A., Needham, S., 1998. Protection of the Environment Near the Ranger Uranium Mine. Report by the Supervising Scientist, Environment Australia.

Koltun, P., 2014. The impact of uranium ore grade on the greenhouse gas footprint of nuclear power. J. Clean. Prod. 84, 360–367. https://doi.org/10.1016/j.jclepro.2013.11.034.

Lenzen, M., 2008. Life-cycle energy balance and greenhouse gas emissions of nuclear energy: a review. Energy Convers. Manag. 49, 2178–2199. https://doi.org/10.1016/j.enconman.2008.01.033.

Lunt, D., Boshoff, P., Boylett, M., El-Ansary, Z., 2007. Uranium extraction: the key process drivers. J. South. African Inst. Min. Metall. 107, 419–426.

Mahmud, M.A.P., Huda, N., Farjana, S.H., Lang, C., 2018. Environmental sustainability assessment of hydropower plant in Europe using life cycle assessment. In: IOP Conference Series: Materials Science and Engineering. https://doi.org/10.1088/1757-899X/351/1/012006.

Moran, C.J., Lodhia, S., Kunz, N.C., Huisingh, D., 2014. Sustainability in mining, minerals and energy: new processes, pathways and human interactions for a cautiously optimistic future. J. Clean. Prod. 84, 1–15. https://doi.org/10.1016/j.jclepro.2014.09.016.

Mudd, G., 2000. Acid in situ leach uranium mining: 1-USA and Australia. Tailings Mine 517–526, 1996.

Mudd, G.M., 2002. Uranium mining in Australia : environmental impact. Radiat. Releases & Rehabil. 179–189.

Mudd, A.G.M., 2006. Uranium Mining: Australia and Globally.

Mudd, G.M., 2008. Radon releases from Australian uranium mining and milling projects: assessing the UNSCEAR approach. J. Environ. Radioact. 99, 288–315. https://doi.org/10.1016/j.jenvrad.2007.08.001.

Mudd, G.M., Diesendorf, M., 2008. Sustainability of uranium mining and milling: toward quantifying resources and eco-efficiency. Environ. Sci. Technol. 42, 2624–2630. https://doi.org/10.1021/es702249v.

Mudd, G.M., Diesendorf, M., 2010. Uranium mining, nuclear power and sustainability: rhetoric versus reality. Sustain. Min. Conf. 39.

Mutchek, M., Cooney, G., Pickenpaugh, G., Marriott, J., Skone, T., 2016. Understanding the contribution of mining and transportation to the total life cycle impacts of coal exported from the United States. Energies 9, 559. https://doi.org/10.3390/en9070559.

Norgate, T., Haque, N., Koltun, P., 2014. The impact of uranium ore grade on the greenhouse gas footprint of nuclear power. J. Clean. Prod. 84, 360–367. https://doi.org/10.1016/j.jclepro.2013.11.034.

Northey, S., Haque, N., Mudd, G., 2013. Using sustainability reporting to assess the environmental footprint of copper mining. J. Clean. Prod. 40, 118–128. https://doi.org/10.1016/j.jclepro.2012.09.027.

Northey, S.A., Mudd, G.M., Saarivuori, E., Wessman-Jääskeläinen, H., Haque, N., 2016. Water footprinting and mining: where are the limitations and opportunities? J. Clean. Prod. 135, 1098–1116. https://doi.org/10.1016/j.jclepro.2016.07.024.

Onn, A.H., Woodley, A., 2014. A discourse analysis on how the sustainability agenda is defined within the mining industry. J. Clean. Prod. 84, 116–127. https://doi.org/10.1016/j.jclepro.2014.03.086.

Parker, D.J., MNaughton, C.S., Sparks, G.A., 2016. Life cycle greenhouse gas emissions from uranium mining and milling in Canada. Environ. Sci. Technol. 50, 9746–9753. https://doi.org/10.1021/acs.est.5b06072.

Parvez Mahmud, M.A., Huda, N., Farjana, S.H., Asadnia, M., Lang, C., 2018. Recent advances in nanogenerator-driven self-powered implantable biomedical devices. Adv. Energy Mater. 8 https://doi.org/10.1002/aenm.201701210.

PE International, 2014. Harmonization of LCA Methodologies for Metals: A Whitepaper Providing Guidance for Conducting LCAs for Metals and Metal Products.

Pellegrino, C., Lodhia, S., 2012. Climate change accounting and the Australian mining industry: exploring the links between corporate disclosure and the generation of legitimacy. J. Clean. Prod. 36, 68−82. https://doi.org/10.1016/j.jclepro.2012.02.022.

Poinssot, C., Bourg, S., Ouvrier, N., Combernoux, N., Rostaing, C., Vargas-Gonzalez, M., Bruno, J., 2014. Assessment of the environmental footprint of nuclear energy systems. Comparison between closed and open fuel cycles. Energy 69, 199−211. https://doi.org/10.1016/j.energy.2014.02.069.

Ranängen, H., Lindman, Å., 2017. A path towards sustainability for the Nordic Mining industry. J. Clean. Prod. 151, 43−52. https://doi.org/10.1016/j.jclepro.2017.03.047.

Raugei, M., Ulgiati, S., 2009. A novel approach to the problem of geographic allocation of environmental impact in Life Cycle Assessment and Material Flow Analysis. Ecol. Indicat. 9, 1257−1264. https://doi.org/10.1016/j.ecolind.2009.04.001.

Santero, N., Hendry, J., 2016. Harmonization of LCA methodologies for the metal and mining industry. Int. J. Life Cycle Assess. 21, 1543−1553. https://doi.org/10.1007/s11367-015-1022-4.

Schmidt, J.H., Thrane, M., 2009. Life Cycle Assessment of Aluminium Production in New Alcoa Smelter in Greenland University 2009.

Schneider, E., Carlsen, B., Tavrides, E., van der Hoeven, C., Phathanapirom, U., 2013a. A top-down assessment of energy, water and land use in uranium mining, milling, and refining. Energy Econ. 40, 911−926. https://doi.org/10.1016/j.eneco.2013.08.006.

Schneider, E., Carlsen, B., Tavrides, E., van der Hoeven, C., Phathanapirom, U., 2013b. Measures of the environmental footprint of the front end of the nuclear fuel cycle. Energy Econ. 40, 898−910. https://doi.org/10.1016/j.eneco.2013.01.002.

Sonter, L.J., Moran, C.J., Barrett, D.J., Soares-Filho, B.S., 2014. Processes of land use change in mining regions. J. Clean. Prod. 84, 494−501. https://doi.org/10.1016/j.jclepro.2014.03.084.

Taylor, G., Farrington, V., Woods, P., Ring, R., Molloy, R., 2004. Review of Environmental Impacts of the Acid In-Situ Leach Uranium Mining Process. CSIRO Land and Water Client Report.

Tost, M., Hitch, M., Chandurkar, V., Moser, P., Feiel, S., 2018. The state of environmental sustainability considerations in mining. J. Clean. Prod. 182, 969−977. https://doi.org/10.1016/j.jclepro.2018.02.051.

Tuusjärvi, M., Mäenpää, I., Vuori, S., Eilu, P., Kihlman, S., Koskela, S., 2014. Metal mining industry in Finland-development scenarios to 2030. J. Clean. Prod. 84, 271−280. https://doi.org/10.1016/j.jclepro.2014.03.038.

Vintró, C., Sanmiquel, L., Freijo, M., 2014. Environmental sustainability in the mining sector: evidence from Catalan companies. J. Clean. Prod. 84, 155−163. https://doi.org/10.1016/j.jclepro.2013.12.069.

Warner, E.S., Heath, G.A., 2012. Life cycle greenhouse gas emissions of nuclear electricity generation: systematic review and harmonization. J. Ind. Ecol. 16, 73−92. https://doi.org/10.1111/j.1530-9290.2012.00472.x.

World Nuclear Association, 2014. Environmental Aspects of Uranium Mining. WNA - World Nuclear Association.

Chapter 5

# Life Cycle Assessment of Copper–Gold–Lead–Silver–Zinc Beneficiation Process

## Introduction

Nonferrous metals like gold, silver, lead, zinc and copper have their unique properties and uses. Copper is a popular conductor of thermal energy and electricity, used commercially for making alloys and building materials. Copper is used for making electronic devices and wiring due to its properties like ductility and softness (Davis, 2001; Norgate and Rankin, 2000). Gold is a popular precious metal for making jewellery, medicine and electronics (Canda et al., 2016; Haque and Norgate, 2014). Like copper, gold is also ductile, soft and bright (Corti and Holliday, 2004; Northey et al., 2014). Lead metals are malleable, making them useful for making bullets, batteries or architectural metals (Christie and Brathwaite, 1995). Due to the high mechanical and electrical conductivity, silver is used for making electronic devices, silverware and jewellery (Lee et al., 2017). Zinc is an alloying metal, abrasive agent and a component to make batteries that have similar chemical properties and applications to lead (Christie and Brathwaite, 1995; Norgate et al., 2007). Though all these materials have their specific properties and applications, they have unique environmental impacts due to the emission of chemical compounds to air, soil, water and environment. Copper is a threatening element for the marine environment and species and is also harmful for deforestation (Ashraf et al., 2015). Gold mining releases a considerable amount of wastes per year which are responsible for soil or water pollution. Underground rock also causes acid mine drainage from gold mining (Fashola et al., 2016). Lead particles can be accumulated in plants or soils which remain unchanged, thus, leading to deforestation (Zuazo and Pleguezuelo, 2008). Silver mining causes the formation of sinkhole, soil or environment pollution or can cause biodiversity (Farjana et al., 2018a). During zinc mining and extraction, its particles emit into the environment and make pollution (Wuana and Okieimen, 2011).

Life Cycle Assessment for Sustainable Mining. https://doi.org/10.1016/B978-0-323-85451-1.00005-6

Among the previous research work based on life cycle assessment (LCA) of nonferrous metal processing, there are a few notable research contributions which addressed the environmental impacts from the mining, beneficiation and refining processes. Many researchers tried to address the LCA analysis of copper, gold or zinc mining (Memary et al., 2012; Norgate and Haque, 2012; Norgate, 2001; Norgate and Rankin, 2000; Qi et al., 2017; Van Genderen et al., 2016). The specific details about these studies are discussed in Table 34 below. Though copper, gold, zinc, lead or silver produce with by-products, their beneficiation process always produces some by-products, whatever the quantity is. However, addressing the allocation of the by-products, their inputs, outputs and emissions is quite different in the previous research works. Memary et al. conducted an LCA analysis of copper production, which had gold and silver metals as a by-product (Memary et al., 2012). Northey et al. conducted their copper-based LCA analysis using weighted data to show relative environmental results in comparison with other metals. Allocation is based on the annual average prices of copper or the London metal exchange rate from 2009 to 2010. It is also mentioned that the nonmetallic by-products included in this study are not weighted for their analysis (Northey et al., 2013). Haque et al. analysed the life cycle environmental impacts of gold with the assumption that gold also contains silver. They used both mass- and revenue-based allocation as the Rio Tinto company employed mass- and revenue-based allocation for concentration and refining (Norgate and Haque, 2012). Genderen et al. analysed the LCA of zinc based on mass-based allocation for allocating the inputs and outputs among the various coproducts during the process (Van Genderen et al., 2016). None of the existing research considered all the base metals for coproduction for environmental impact analysis from the beneficiation process. In this research, particularly, the beneficiation process is considered with the coproducts where environmental effects are assessed through 15 impact categories. In the previous works, mostly the global warming potential, acidification potential and energy consumption were analysed. Their lack of focus persists in many significant impact categories like ecotoxicity, eutrophication or ionising radiation. These limitations of previous studies are also covered here. Moreover, there was no significant LCA research based on lead or silver production processes, so this is another notable aspect of the present work. The key features of the existing LCA work on copper, gold or zinc production processes are described in Table 34 below.

The reasons behind these environmental issues vary from one mining step to another. From extraction, development, mining, beneficiation, refinery to tailing disposal, each step has separate heat and energy requirements and different technologies. These arrangements make them act differently towards the environment. To assess the environmental impacts caused by different stages, it is necessary to calculate the life cycle environmental impacts on the beneficiation stage of mining. Life cycle environmental impact analysis is a powerful tool which aims to analyse the details of environmental effects from

**TABLE 34** Key features of the existing research papers on life cycle assessment of copper, gold and zinc.

| Metal | Author | Objectives | Impact categories | Responsible process steps/materials | Analysis results |
|-------|--------|-----------|-------------------|-------------------------------------|------------------|
| Copper | Norgate (Norgate and Rankin, 2000) | Assessed the pyrometallurgical and hydrometallurgical processing routes, number of process variables and ore grades. | Energy consumption, greenhouse gas emission and acidification potential. | Hydrometallurgical production has a higher impact on the environment. | GWP 3.3 $CO_2$ eq/kg. Energy consumption 33 MJ/kg. Acidification potential 0.04 kg $SO_2$ eq/kg. |
| | Memary et al. (Memary et al., 2012) | Analysis of copper production based on time series in Australia from 1940 to 2008 (5 largest mines). | Photochemical ozone depletion potential, global warming potential and acidification potential. | Ore grade, differences in technology and regional energy sources impact mainly on environmental effects. | Carbon footprint varies from 2.5 to 8.5 kg $CO_2$ eq/kg. |
| | Norgate (Norgate, 2001) | Life cycle analysis of different copper mining technologies using hydrometallurgical processes. | Energy consumption, greenhouse gas emission and acidification potential. | Hydrometallurgical processes involving the electrowinning processes have the most substantial energy consumption due to power consumption. | From 4.3 to 8.9 kg $CO_2$ eq/kg for hydrometallurgical processes. From 1.5 to 4.2 kg $CO_2$ eq/kg for pyrometallurgical processes. |
| | Northey et al. (Northey et al., 2013) | Quantified the environmental footprint caused by the copper production processes using sustainability reporting. | Energy consumption, greenhouse gas emission and water intensity. | The variation of results from one company to another is due to copper mining, ore grade, the source of energy and reporting procedure of companies, which should be clarified in the company reports. | From 1 to 9 t $CO_2$ eq/Cu greenhouse gas emissions and around 70.4 kL/t Cu water footprint. |

*Continued*

**TABLE 34** Key features of the existing research papers on life cycle assessment of copper, gold and zinc.—cont'd

| Metal | Author | Objectives | Impact categories | Responsible process steps/materials | Analysis results |
|---|---|---|---|---|---|
| Gold | Chen et al. (Chen et al., 2018) | Assessed the environmental impacts of the gold production of China on ecosystems and human health. | Several major impact categories are assessed through ReCiPe method. | Major impacts are due to ore mining and energy consumption. | Climate change effect is 5.55E7 kg $CO_2$ eq. A major part of the impact is on the metal depletion category. |
| | Norgate et al. (Norgate and Haque, 2012) | Life cycle environmental impact assessment of gold production for two types of ores has been conducted. | Embodied energy, greenhouse gas footprint is assessed. | With refractory ore, the emissions are 50% higher than those with nonrefractory ores due to excess material and technology use. Mining and comminution stage effects are mostly due to electricity consumption. | 200000 GJ/t Au. 18000 t $CO_2$ eq/t Au. 260000 t water/t Au. 1,270,000 t waste/t Au. |
| Zinc | Genderen et al. (Van Genderen et al., 2016) | Life cycle analysis of zinc production has been conducted while datasets are from 24 mines and 18 smelters. | Global warming, acidification, eutrophication, photochemical ozone creation and primary energy demand are assessed. | 65% of environmental burdens from global zinc production are due to smelting, 30% is from mining and 5% is from transportation. | Primary energy demand is 37500 MJ/t of Zn. Climate change effect is 2600 kg $CO_2$ eq/t. |
| | Qi et al. (Chen et al., 2018) | Life cycle analysis of zinc production in China is conducted while datasets are from national statistical dataset inventory and process-based life cycle inventory. | Assessed several impact categories from ReCiPe method, including climate change, human toxicity and freshwater ecotoxicity. | A major contributor is zinc ore mining and energy consumption. | Climate change 6.12E3 kg $CO_2$ eq. Marine ecotoxicity 17 kg 1,4-DB eq. Metal depletion 3.58E3 kg Fe eq. |

cradle-to-gate of a product, system or process. In this research, we are focussed on the sustainability assessment of the beneficiation process of gold−silver−lead−zinc−copper mines.

## Copper-Gold-Lead-Silver-Zinc Beneficiation Process

Beneficiation includes crushing, grinding, gravity concentration and flotation concentration. Beneficiation is followed by processing activities such as smelting and refining. The beneficiation process begins with milling, which is followed by flotation for further beneficiation. At the first stage, extracted ores undergo the milling operation to produce uniformly sized particles for crushing, grinding, wet or dry concentration. The type of milling operable in a certain plant is chosen by capital investment and economics. The degree of crushing or grinding, which is required for further beneficiation, is dependent on capital. Crushing is a dry operation which only involves dust control using water spray (Drzymala, 2007). A primary or jaw crusher is located at the mine site and reduces the particle diameter of the ores into <6 in. The crushed ore is then transported to the mill site for crushing, grinding, classification and concentration. The second stage, grinding, is a wet operation which requires initial flotation and water to make a slurry. The hydrocyclone operates between each grinding operation to classify the type of particles: fine or coarse (Long et al., 1998).

This process is used to adhere to ore mineral or a group of minerals with the air bubbles after involving chemical reagents in operation. Chemical reagents got reacted with the desired mineral in the flotation process. The effectiveness of the flotation technique is dependent on four factors: the degree of oxidation of the ore, the number of copper minerals present, the nature of the gangue and the presence of iron sulphides. There are some other important factors such as the particle size, minerals compatible with the reagents and the condition of the water. Conditioners and regulators might be used during or after the milling time for ore treatment (Drzymala, 2007). Flotation is an effective method to concentrate the targeted elements existed in minerals based on the difference in physicochemical properties of various mineral surfaces. It can easily separate copper (Feng et al., 2018b), lead (Feng et al., 2017a), zinc (Feng and Wen, 2017) and tin (Feng et al., 2017b) minerals from gangue minerals by addition of flotation reagents. The concentrates of minerals must go through pyrometallurgical methods like smelting and refining. However, before these steps, the concentrates may require roasting and sintering, which depends on the processing method. The ore concentrate undergoes partial fusion which turns it into agglomerated material suitable for processing operations (Drzymala, 2007). The sintering operation consists of blending, sintering, cooling and sizing. At first, the raw material concentrates are blended with moistures in mills, drums or pans. This step is called blending. In the next step, the concentrate feed is fired or sintered and then cooled (Long et al., 1998). The sinter gets crushed with being cool. Then the concentrate will be graded. After grading, it is crushed to produce a smaller sinter product. In roasting, gas—solid reactions are involved at elevated temperatures, which purify the metal by treating it with hot air (Shedd, 2016).

## Life Cycle Assessment

Life cycle assessment is an internationally recognised standardised process based on ISO 14040. ISO 14040 is composed of four major steps of impact analysis: goal and scope definition, life cycle inventory analysis, life cycle impact assessment and interpretation of the results (Awuah-Offei and Adekpedjou, 2011; Lima et al., 2018; Mahmud et al., 2018a, 2018b, 2018c, 2018d; Zhang et al., 2018).

The goal of this chapter is to analyse the cradle-to-gate environmental effects of the gold—silver—lead—zinc—copper beneficiation process, thus comparing their impacts under several impact categories. The scope of this chapter is the environmental emissions which are generated or emitted from the milling operation, flotation, roasting and sintering. The materials, energy (renewables and nonrenewables), fossil fuels used for process heat generation, organic and inorganic chemicals are considered as material inputs or inputs from nature. On the other hand, the emissions to soil, water (groundwater or other water resources), air and the final production wastes are considered as outputs of the beneficiation process (Farjana et al., 2018b). Table 35 shows the life cycle inventory inputs and outputs which include fuels, renewable and nonrenewable energies used for the beneficiation process and electricity generation, materials, organic and inorganic chemicals, emissions to air, soil, water (groundwater resources and other forms of water resources) and final waste emissions. For this analysis, 1 kg of each of the coproducts of each metal is considered here to be produced from the beneficiation process, that is the purified ore at the end of the mine concentrating stage. LCA of the nonferrous metal beneficiation processes includes ore from the mine, the ore concentration, transportation and smelting. This study does not have any intention to include the end-of-life product stages or environmental emissions for these nonferrous metals. A cradle-to-gate LCA analysis has been performed, which is focussed particularly on the metal beneficiation process. Fig. 22 describes the system boundary followed in the present research for LCA analysis. Fig. 23 shows the material flows from the nonferrous metal beneficiation process. In the third stage of life cycle analysis, life cycle impact analysis is carried out using SimaPro software version 8.5. There are few other standard software available for LCA analysis such as Gabi, OpenLCA and Umberto. However, SimaPro is very well resourced with latest datasets and most widely used for conducting life cycle impact analysis (LCIA) focussed on metal mining industries. SimaPro software provides easy access and integrity with renowned and validated databases like EcoInvent, USGS

TABLE 5.5 Life Cycle Inventory datasets — inputs and outputs.

| Inputs | Copper | Gold | Lead | Silver | Zinc | Unit |
|---|---|---|---|---|---|---|
| Zn 0.63%, Au 9.7E-4%, Ag 9.7E-4%, Cu 0.38% and Pb 0.014%, in ore, in ground | 1.1 | 1.1 | 1.1 | 1.1 | 1.1 | Kg |
| Water, salt and ocean | 0.019 | 68.992 | 0.005 | 1.178 | 0.008 | m³ |
| Water, well, in ground | 0.001 | 5.411 | 0.0004 | 0.092 | 0.0006 | m³ |
| Water, river | 0.008 | 31.113 | 0.002 | 0.531 | 0.0037 | m³ |
| Electricity, medium voltage | 3.219 | 11289.23 | 0.883 | 192.788 | 1.35 | kWh |
| Hard coal | 1.77 | 6209.009 | 0.485 | 106.032 | 0.742 | MJ |
| Diesel | 0.965 | 3386.75 | 0.264 | 57.836 | 0.405 | MJ |
| Natural gas | 0.482 | 1693.375 | 0.132 | 28.918 | 0.202 | MJ |
| Heavy fuel oil | 0.643 | 2257.768 | 0.176 | 38.556 | 0.27 | MJ |
| Blasting | 0.115 | 406.48 | 0.031 | 6.941 | 0.048 | kg |
| Transport, lorry >16t | 0.013 | 46.249 | 0.003 | 0.789 | 0.005 | tkm |
| Transport, freight | 0.107 | 378.217 | 0.029 | 6.458 | 0.045 | tkm |
| Mine, gold–silver–zinc–lead–copper | 1.12E-10 | 3.94E-07 | 3.08E-11 | 6.72E-09 | 4.71E-11 | p |
| Tap water | 3E-04 | 1.352 | 1E-04 | 0.023 | 1.6E-04 | kg |
| Sodium cyanide | | 27.961 | | 0.47 | | kg |
| Limestone | | 62.559 | | 1.053 | | kg |
| Sodium hydroxide, 50% in $H_2O$ | | 11.524 | | 0.194 | | kg |
| Zinc, primary | | 2.864 | | 0.048 | | kg |
| Charcoal | | 16.791 | | 0.282 | | kg |
| Sulphuric acid, liquid | | 0.658 | | 0.011 | | kg |
| Hydrochloric acid, 30% in $H_2O$ | | 13.499 | | 0.227 | | kg |

| Outputs | Copper | Gold | Lead | Silver | Zinc | Unit |
|---|---|---|---|---|---|---|
| Copper | 1.04E-07 | 3E-04 | 2.85E-08 | 6.23E-06 | 4.36E-08 | kg |
| Carbon dioxide, fossil | 0.018 | 65.036 | 0.005 | 1.11 | 0.007 | kg |
| Lead | 3.38E-08 | 1E-04 | 9.28E-09 | 2.02E-06 | 1.42E-08 | kg |
| Zinc | 2.87E-08 | 1E-04 | 7.88E-09 | 1.72E-06 | 1.2E-08 | kg |
| Heat, waste | 11.591 | 40,640.21 | 3.179 | 694.019 | 4.861 | MJ |
| Copper, ion | 1.42E-06 | 0.004 | 3.9E-07 | 8.5E-05 | 5.96E-07 | kg |
| Lead | 7.62E-07 | 0.002 | 2.09E-07 | 4.56E-05 | 3.2E-07 | kg |
| Zinc, ion | 1.21E-05 | 0.042 | 3.33E-06 | 7E-04 | 5.09E-06 | kg |
| Disposal, sulphidic tailings | 4.087 | 14332.72 | 1.137 | 1.137 | 1.137 | Kg |

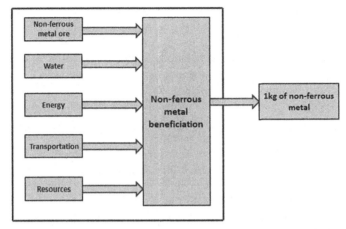

**FIGURE 22** System boundary diagram for life cycle assessment analysis.

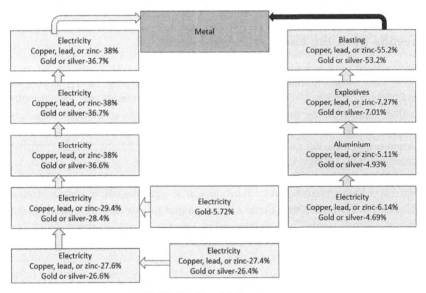

**FIGURE 23** Material flow diagram.

and AusLCI, which contain numerous datasets of mining and mineral processing industries. This feature is limited to other available software. However, EcoInvent can be purchased separately and integrated with any software in use, like for Gabi. The relevant datasets originated from databases and literature that are considered here is from 'Lifecycle inventories for metals and methodological approaches' by Classen et al. (Althaus and Classen, 2005).

The geographical coverage for the original dataset is Sweden; however, for the ease of analysis, it is assumed that geographical or climate-related factors are negligible here. Also, it is hard to compare among the datasets used in the present research and the previous researches because most of them considered the whole mining process from extraction to refining as a unit process (Farjana et al., 2018a; Mahmud et al., 2018a, 2018b, 2018c, 2018d). There is no research specifically focussed on the beneficiation process. Also, a lack of research focus existed in the LCA of the silver or lead mining processes. However, in the later stages of this research, sensitivity analysis based on the electricity grid mix of different power associations in Europe and the energy mix of fossil fuels is presented to justify the impact of the differences of energy on gold−silver−lead−zinc−copper beneficiation process. Gold, silver, lead, zinc and copper − all the nonferrous metal coproducts − are considered in this analysis. The revenue-based allocation approach is followed here, to distribute the process material inputs and outputs, emissions, wastes and other flows associated with this process. The details of the allocation techniques can be found in the SimaPro manual (Goedkoop et al., 2014).

## Results from the Life Cycle Assessment

Table 36 and Fig. 24 describe the comparative LCA results among the five nonferrous metals involved in the couple production: gold, silver, lead, zinc and copper. The results indicate that among the five metals, the gold beneficiation process is the most detrimental one towards sustainability. The significant impact categories are ionising radiation, photochemical ozone formation, terrestrial eutrophication, acidification, climate change, marine eutrophication, freshwater eutrophication and water resources depletion. Blasting is the most environmentally impactful process of the beneficiation process, which impacts mostly onto terrestrial eutrophication, marine eutrophication and acidification. Electricity consumed in copper beneficiation is the second largest impactful element which mostly affects ionising radiation. Carbon-14 emission from medium voltage electricity generation is responsible for ionising radiation.

On the other hand, blasting releases nitrogen oxides, which cause terrestrial eutrophication, marine eutrophication and acidification. The most significant environmental impact is on ionising radiation, which is 2148.9 kBq U235 eq from gold beneficiation. The next one is terrestrial eutrophication which is 172.22 kg NMVOC eq. The photochemical ozone formation effect is 3335.388 kg C deficit. The acidification potential is 0.019 CTUe. The human toxicity noncancer effects are counted as 3.88E-04 CTUh. From the gold beneficiation, the human toxicity cancer effects are counted as 2.37E-05 CTUh. The impacts from marine eutrophication are 951.064 molc N eq. Lastly, the particulate matter effect is 10.697 kg PM2.5 eq.

TABLE 36 Comparative life cycle assessment results of beneficiation process – International Reference Life Cycle Data System method.

| Label | Unit | Copper | Gold | Lead | Silver | Zinc |
|---|---|---|---|---|---|---|
| CC (climate change) | kg CO$_2$ eq | 0.97 | 3640.55 | 0.268 | 62.12 | 0.41 |
| OD (ozone depletion) | kg CFC-11 eq | 6.06E-08 | 2.25E-04 | 1.66E-08 | 3.84E-06 | 2.54E-08 |
| HTNCE (human toxicity, noncancer effects) | CTUh | 7.17E-08 | 3.88E-04 | 1.97E-08 | 6.59E-06 | 3.01E-08 |
| HTCE (human toxicity, cancer effects) | CTUh | 6.37E-09 | 2.37E-05 | 1.75E-09 | 4.05E-07 | 2.67E-09 |
| PM (particulate matter) | kg PM2.5 eq | 3E-03 | 10.697 | 8.3E-04 | 0.18 | 1.2E-03 |
| IRHH (ionising radiation HH) | kBq U235 eq | 0.613 | 2148.9 | 0.168 | 36.697 | 0.257 |
| AP (acidification) | CTUe | 5.53E-06 | 0.019 | 1.52E-06 | 3.3E-04 | 2.32E-06 |
| TE (terrestrial eutrophication) | kg NMVOC eq | 0.048 | 172.22 | 0.013 | 2.94 | 0.02 |
| FE (freshwater eutrophication) | molc H+ eq | 0.054 | 191.714 | 0.014 | 3.273 | 0.022 |
| ME (marine eutrophication) | molc N eq | 0.27 | 951.064 | 0.074 | 16.24 | 0.113 |
| FE (freshwater ecotoxicity) | kg P eq | 8.08E-04 | 2.84 | 2.22E-04 | 0.048 | 3.39E-04 |
| LU (land use) | kg N eq | 0.016 | 59.25 | 0.004 | 1.011 | 0.007 |
| WRD (water resource depletion) | CTUe | 0.292 | 1193.144 | 0.08 | 20.33 | 0.122 |
| POF (photochemical ozone formation) | kg C deficit | 0.438 | 3335.388 | 0.12 | 56.539 | 0.183 |
| MFRRD (mineral, fossil and renewable resource depletion) | m$^3$ water eq | 0.005 | 19.882 | 1.4E-04 | 0.339 | 2.2E-03 |

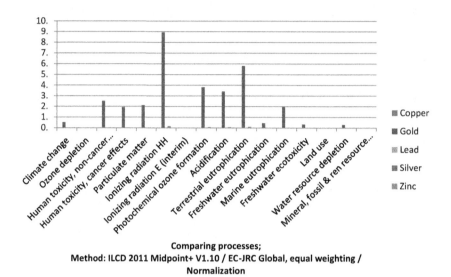

Comparing processes;
Method: ILCD 2011 Midpoint+ V1.10 / EC-JRC Global, equal weighting /
Normalization

**FIGURE 24** Analysis results from international reference life cycle data system method — normalised results (normalised results).

On the other hand, lead beneficiation is the most sustainable to the environment. The most significant environmental impact is on ionising radiation, which is 0.168 kBq U235 eq from lead beneficiation. The next one is terrestrial eutrophication which is 0.013 kg NMVOC eq. The photochemical ozone formation effect is 0.12 kg C deficit. The acidification potential is 1.52E-06 CTUe. The human toxicity noncancer effects are counted as 1.97E-08 CTUh. From the lead beneficiation, the human toxicity cancer effects are counted as 1.75E-09 CTUh. The impacts from marine eutrophication are 0.074 molc N eq. Lastly, the particulate matter effect is 8.3E-04 kg PM2.5 eq. Environmental impacts from silver beneficiation come next to gold, while copper and zinc beneficiation show similar results following silver beneficiation. Table 37 shows the characterised results, while Fig. 24 shows the normalised results to compare the impacts caused by the coproduct metals.

Table 37 and Fig. 25 describe the LCIA results from the cumulative energy demand (CED) method. The analysis results from the CED method show that the highest environmental effects are caused by nuclear energy consumption and coal consumption which might have been due to electricity generation. Analysis results showed that gold beneficiation consumed the highest amount of nuclear fuel that is 54,965.26 MJ LHV. Then renewable energy consumption is 17,551.69 MJ LHV. Embodied energy consumption is 120,798.3 MJ LHV and 123,234.2 MJ HHV. Coal consumption generates 15,431 MJ LHV heat, while the smallest amount of heat is from gas that is 8903 MJ LHV. The CED method shows similar results to the International Reference Life Cycle

**TABLE 37** Comparative life cycle assessment results of beneficiation process — Cumulative Energy Demand method.

| Impact category | Unit | Copper | Gold | Lead | Silver | Zinc |
|---|---|---|---|---|---|---|
| Renewables | MJ LHV | 4.988 | 17,551.69 | 1.368 | 299.719 | 2.092 |
| Fossil fuels — oil | MJ LHV | 3.365 | 12,190.47 | 0.923 | 208.087 | 1.411 |
| Fossil fuels — gas | MJ LHV | 2.064 | 8903.549 | 0.566 | 151.659 | 0.865 |
| Fossil fuels — coal | MJ LHV | 4.099 | 15,431 | 1.124 | 263.271 | 1.719 |
| Biomass | MJ LHV | 2.934 | 11,756.8 | 0.804 | 200.43 | 1.23 |
| Nuclear | MJ LHV | 15.66 | 54,965.26 | 4.295 | 938.637 | 6.567 |
| Embodied energy LHV | MJ LHV | 33.112 | 120,798.3 | 9.082 | 2061.798 | 13.88 |
| Embodied energy HHV | MJ HHV | 33.709 | 123,234.2 | 9.246 | 2103.316 | 14.13 |

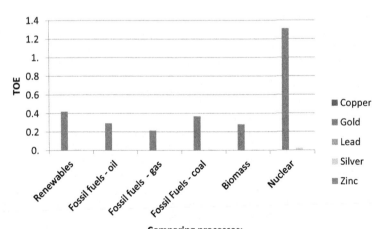

Comparing processes;
Method: Cumulative Energy Demand - by energy type/ Energy by fuel source / Weighting

**FIGURE 25** Analysis results from cumulative energy demand method (weighted results).

Data System (ILCD) method that lead beneficiation consumes the lowest amount of energy. Lead beneficiation consumed the lowest amount of nuclear fuel that is 4.295 MJ LHV. Then renewable energy consumption is 1.368 MJ LHV. Embodied energy consumption is 9.082 MJ LHV and 9.2 MJ HHV. Coal consumption generates 1.124 MJ LHV of heat, while the smallest amount of heat is from gas that is 0.566 MJ LHV. Silver beneficiation has significant energy consumption, less than for gold beneficiation and greater than for copper. Copper and zinc beneficiation show similar characteristics.

The IMPACT 2002+ method analyses the life cycle impacts of a product, process or a system of processes based on endpoint indicator-based categories. The endpoint indicator-based categories are human health, ecosystems quality, climate change and resources. The human health impacts from the gold beneficiation are 0.018758 DALY, whereas from lead beneficiation it is 1.46E-06 DALY. The ecosystems quality impacts from the gold beneficiation are 64,910.84 $PDF * m^2 * yr$, whereas from lead beneficiation it is 5.064 $PDF * m^2 * yr$. The resources impacts from gold beneficiation are 8.59E-03 MJ primary, whereas from lead beneficiation it is 6.61E-07 MJ primary. The climate change impacts from gold beneficiation are 3501.226 kg $CO_2$ eq, whereas from the lead beneficiation it is 0.256 kg $CO_2$ eq. The comparative analysis based on this method shows here that among the five metals, the gold beneficiation process predominates on the endpoint indicator-based categories. The lead beneficiation process shows the greatest sustainability among the coproduct metals. The analysis results are presented in Table 38 and Fig. 26.

Previous research studies that assessed the environmental impacts caused by the gold mining processes showed that due to a lower ore grade, gold mining consumes a huge amount of electricity and thus has large greenhouse

**TABLE 38** Comparative life cycle assessment results of beneficiation process — IMPACT 2002+ method.

| Damage category | Unit | Copper | Gold | Lead | Silver | Zinc |
| --- | --- | --- | --- | --- | --- | --- |
| Human health | DALY | 5.32E-06 | 0.018 | 1.46E-06 | 3.2E-04 | 2.23E-06 |
| Ecosystem quality | PDF × m$^2$ × yr | 18.463 | 64910.84 | 5.064 | 1108.453 | 7.743 |
| Climate change | kg CO$_2$ eq | 0.935 | 3501.226 | 0.256 | 59.739 | 0.392 |
| Resources | MJ primary | 2.41E-06 | 8.59E-03 | 6.61E-07 | 1.47E-04 | 1.01E-06 |

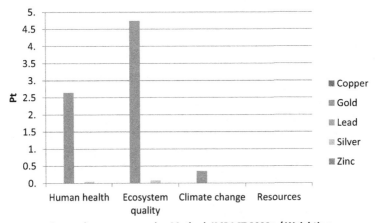

**Comparing processes using Method: IMPACT 2002+ / Weighting**

**FIGURE 26** Analysis results from the IMPACT 2002+ method (weighted results).

gas (GHG) effects (Chen et al., 2018; Haque and Norgate, 2014; Norgate and Haque, 2012). The major contribution of the present research is that this research specifically focusses on the metal beneficiation process, whereas most of the existing studies considered the whole mining process as a unit process. It is very crucial to identify the energy-intensive or environmentally detrimental process to assess the renewable energy integration potential in a process (Farjana et al., 2018c). Also, this work considers major impact categories, which lacks focus in previous work, mostly based on GHG emissions, energy demand and acidification. Of course, those are very important impact categories, but the greater environmental effects from these nonferrous metals are on the ionising radiation field, which is a notable finding from this work. Table 39 shows the detailed life cycle emissions which are responsible for affecting a specific impact category in the combined metal beneficiation process. For climate change, carbon dioxide biogenic and carbon dioxide fossil are responsible for the highest emissions. Ethane, 1,2-dichloro-1,1,2,2-tetrafluoro- and CFC-114 emission are responsible for causing ozone depletion. Similarly, mercury is for human toxicity noncancer effects, and chromium is for human toxicity cancer effects. Ammonia and nitrogen oxides are affecting particulate matter, acidification potential, metal depletion and terrestrial eutrophication. Carbon and cesium emissions cause ionising radiation effects on human health and ecosystems. Phosphate is affecting freshwater ecotoxicity. In comparison among the coproduct metals, gold mining emits the largest amount of carbon dioxide, ammonia, nitrogen dioxides and chromium. On the other hand, lead beneficiation shows the greatest sustainability among all the coproduct metals.

**TABLE 39** Life cycle inventory emissions responsible for causing environmental effects per impact categories.

| Impact category | Substance | Unit | Copper | Gold | Lead | Silver | Zinc |
|---|---|---|---|---|---|---|---|
| CC | Carbon dioxide, biogenic | kg CO₂ eq | 0.279328 | 983.1072 | 0.076617 | 16.78781 | 0.117146 |
| | Carbon dioxide, fossil | kg CO₂ eq | 0.236714 | 937.5583 | 0.06493 | 15.98579 | 0.099277 |
| OD | Ethane, 1,2-dichloro-1,1,2,2-tetrafluoro- and CFC-114 | kg CFC-11 eq | 2.87E-08 | 0.000101 | 7.88E-09 | 1.72E-06 | 1.2E-08 |
| HTNCE | Methane, bromotrifluoro- and halon 1301 | kg CFC-11 eq | 3.14E-08 | 0.000112 | 8.61E-09 | 1.9E-06 | 1.32E-08 |
| | Mercury | CTUh | 5.74E-09 | 4.01E-05 | 1.57E-09 | 6.8E-07 | 2.41E-09 |
| HTCE | Mercury | CTUh | 5.29E-09 | 2.84E-05 | 1.45E-09 | 4.82E-07 | 2.22E-09 |
| | Chromium VI | CTUh | 8.42E-10 | 2.95E-06 | 2.31E-10 | 5.05E-08 | 3.53E-10 |
| | Chromium VI, ground | CTUh | 4.76E-09 | 1.77E-05 | 1.31E-09 | 3.02E-07 | 2E-09 |
| PM | Ammonia | kg PM2.5 eq | 0.000456 | 1.600453 | 0.000125 | 0.027331 | 0.000191 |
| | Nitrogen oxides | kg PM2.5 eq | 0.000288 | 1.012937 | 7.9E-05 | 0.017297 | 0.000121 |
| IRHH | Carbon-14 | kBq U235 eq | 0.577562 | 2025.006 | 0.15842 | 34.58137 | 0.242221 |
| IRE | Carbon-14 | CTUe | 4.09E-06 | 0.014337 | 1.12E-06 | 0.000245 | 1.71E-06 |
| | Cesium-137 | CTUe | 1.13E-06 | 0.003952 | 3.09E-07 | 6.75E-05 | 4.73E-07 |
| POF | Nitrogen oxides | kg NMVOC eq | 0.039907 | 140.296 | 0.010947 | 2.395772 | 0.016737 |
| | NMVOC, nonmethane volatile organic compounds | kg NMVOC eq | 0.007103 | 24.92484 | 0.001948 | 0.425641 | 0.002979 |
| AP | Ammonia | molc H+ eq | 0.020663 | 72.4643 | 0.005668 | 1.237482 | 0.008666 |

| | | | | | | | |
|---|---|---|---|---|---|---|---|
| | Nitrogen oxides | molc H+ eq | 0.029531 | 103.8191 | 0.0081 | 1.772871 | 0.012385 |
| TE | Ammonia | molc N eq | 0.092368 | 323.9298 | 0.025337 | 5.531789 | 0.038739 |
| | Nitrogen oxides | molc N eq | 0.170003 | 597.6611 | 0.046632 | 10.20599 | 0.071299 |
| FE | Phosphate | kg P eq | 0.000803 | 2.821253 | 0.00022 | 0.048178 | 0.000337 |
| ME | Nitrogen oxides | kg N eq | 0.015524 | 54.57516 | 0.004258 | 0.931955 | 0.006511 |
| FE(Ecotox) | Antimony | CTUe | 0.004578 | 62.8252 | 0.001256 | 1.061984 | 0.00192 |
| | Chlorothalonil | CTUe | 0.023869 | 83.74461 | 0.006547 | 1.430108 | 0.010011 |
| LU | Occupation, dump site | kg C deficit | 0.074997 | 263.6401 | 0.020571 | 4.502064 | 0.031453 |
| | Occupation, forest, extensive | kg C deficit | 0.33132 | 2949.001 | 0.090879 | 49.94369 | 0.138951 |
| WRD | Water, cooling, unspecified natural origin/m³ | m³ water eq | 0.001448 | 6.017579 | 0.000397 | 0.102544 | 0.000607 |
| | Water, river | m³ water eq | 0.003435 | 12.08332 | 0.000942 | 0.20634 | 0.001441 |
| MERED | Gas, mine, off-gas, process and coal mining/m³ | kg Sb eq | 1.88E-14 | 6.69E-11 | 5.15E-15 | 1.14E-12 | 7.87E-15 |

## Sensitivity Analysis based on Electricity Mix and Energy Mix

Table 40 shows the sensitivity analysis results particularly focussed on the beneficiation of gold. Five different case scenarios are considered here:

Grid mix 1 — the base case with medium voltage electricity for Sweden.

Grid mix 2 — the modified case with medium voltage electricity for CENTRAL (Central European Power Association).

Grid mix 3 — the modified case with medium voltage electricity for NORDEL (Nordic Countries Power Association).

Grid mix 4 — the modified case with medium voltage electricity for RER (Europe).

Grid mix 5 — the modified case with 50% reduction of hard coal, 50% reduction of diesel fuel and replacement of the equivalent amount of energy supply by natural gas.

Among the analysis results presented here, scenarios 2, 3 and 4 show a significant increase in climate change but nearly unchanged results of terrestrial eutrophication, ozone depletion and freshwater eutrophication but surprisingly, much reduced impacts in the fields of ionising radiation and human toxicity. On the other hand, scenario 5, which shows the effects of reducing 50% of coal consumption and 50% of diesel consumption and replacing the heat produced by them with natural gas, shows that the climate change, eutrophication (terrestrial and freshwater) and acidification effects will be reduced a little bit with no significant change in ionising radiation. It is clear from the sensitivity analysis results that electricity consumption is the dominant factor for gold beneficiation. The sensitivity analysis conducted here reveals three basic facts. The first is the applicability of the dataset if the same technological mining principles are used but in the different regions with separate grid mix in European countries. The results from scenarios 2, 3 and 4 show that impact analysis results are similar within the medium voltage electricity grid mixes within Europe, even though the life cycle inventory datasets are particularly based on mines in Sweden. Another important fact from scenarios 2, 3 and 4r is that a modification of electricity grid mix acts oppositely among ionising radiation and climate change. Even though climate change effects increase in some cases, ionising radiation decreases. Thirdly, scenario 5 shows the energy mix scenario by replacing coal and diesel with natural gas. This case shows the minimal change in environmental impacts can be caused by fossil fuel replacement since electricity consumption is not reduced, which is the most crucial factor for the nonferrous metal beneficiation process. In the case of metals assessed here other than gold, there are few types of the research reported in the literature which tried to attempt a life cycle impact analysis of copper or zinc production processes but none in the open

**TABLE 40** Sensitivity analysis results of gold beneficiation.

| Impact category/gold | Grid mix 1 – Sweden | Grid mix 2 – CENTRAL | Grid mix 3 – NORDEL | Grid mix 4 – RER | Grid mix 5 – energy mix | Unit |
|---|---|---|---|---|---|---|
| CC (climate change) | 3640.552 | 12,617.75 | 14,056.23 | 8142.061 | 3473.757 | kg CO$_2$ eq |
| OD (ozone depletion) | 0.00022 | 0.00018 | 0.0001 | 0.0002 | 0.0002 | kg CFC-11 eq |
| HTNCE (human toxicity, noncancer effects) | 0.00038 | 0.00072 | 0.0003 | 0.0004 | 0.0003 | CTUh |
| HTCE (human toxicity, cancer effects) | 2.37E-05 | 0.000118 | 2.36E-05 | 4.02E-05 | 2.29E-05 | CTUh |
| PM (particulate matter) | 10.697 | 10.624 | 7.944 | 9.38 | 10.179 | kg PM2.5 eq |
| IIHH (ionising radiation HH) | 2148.905 | 852.482 | 0.289 | 1452.374 | 2148.86 | kBq U235 eq |
| IIE (ionising radiation E) | 0.0194 | 0.00769 | 1.67E-06 | 0.013 | 0.0194 | CTUe |
| POF (photochemical ozone formation) | 172.22 | 188.807 | 198.357 | 180.182 | 169.3 | kg NMVOC eq |
| AP (acidification) | 191.714 | 245.336 | 208.237 | 215.203 | 187.484 | molc H+ eq |
| TE (terrestrial eutrophication) | 951.064 | 1008.62 | 1058.83 | 978.379 | 940.453 | molc N eq |
| FE (freshwater eutrophication) | 2.844 | 2.833 | 2.841 | 2.834 | 2.844 | kg P eq |
| ME (marine eutrophication) | 59.255 | 64.556 | 69.298 | 61.78 | 58.28 | kg N eq |
| FE (freshwater ecotoxicity) | 1193.144 | 2731.087 | 1147.639 | 1656.817 | 1165.771 | CTUe |
| LU (land use) | 3335.388 | 3041.193 | 2283.141 | 2913.374 | 3313.839 | kg C deficit |
| WRD (water resource depletion) | 19.88 | 65.699 | 11.752 | 41.269 | 19.82 | m$^3$ water eq |
| MFRRD (mineral, fossil and renewable resource depletion) | 6.69E-11 | 5.67E-11 | 4.95E-11 | 1.49E-10 | 6.63E-11 | kg Sb eq |

literature about lead or silver production (Memary et al., 2012; Norgate et al., 2007; Norgate and Rankin, 2000; Qi et al., 2017). Previous researches on copper or zinc lack significant impact categories like ionising radiation, eutrophication, ecotoxicity and human toxicity.

## Discussion

Table 41 compares among the GHG emission results from the previous studies on LCA of mining (copper, gold, lead, silver and zinc) and this research. However, the research scope, system boundary, analysis method, geographic region consideration, allocation technique, energy mix and heat mix vary among the studies, even though the studies are conducted for the same metal. The results are compared only concerning GHG because most of the previous studies did not conduct a full LCA, which makes it harder to compare for all the major environmental impact categories. The comparison shows that for copper mining, the GHG emission estimation ranges between 1.5 and 8.9 kg $CO_2$ eq/kg depending on those factors outlined above. From this study, the calculated GHG emission result is 0.97 kg $CO_2$ eq., which agrees well with

**TABLE 41** Comparison of results with previous studies.

| Metal | Study | GHG emission results |
|---|---|---|
| Copper | (Norgate and Rankin, 2000) | 3.3 kg $CO_2$ eq/kg |
| | (Memary et al., 2012) | 2.5–8.5 kg $CO_2$ eq/kg |
| | (Norgate, 2001) | 4.3–8.9 kg $CO_2$ eq/kg 1.5–4.2 kg $CO_2$ eq/kg |
| | (Northey et al., 2013) | 1 to 9 t $CO_2$ eq/Cu |
| | This research | 0.97 kg $CO_2$ eq |
| Gold | (Chen et al., 2018) | 5.55E7 kg $CO_2$ eq |
| | (Norgate and Haque, 2012) | 18000 t $CO_2$ eq/t Au |
| | This research | 3640.55 kg $CO_2$ eq |
| Zinc | (Van Genderen et al., 2016) | 2600 kg $CO_2$ eq/t |
| | (Chen et al., 2018) | 6.12E3 kg $CO_2$ eq |
| | This research | 0.41 kg $CO_2$ eq |
| Silver | This research | 62.12 kg $CO_2$ eq |
| Lead | This research | 0.268 kg $CO_2$ eq |

the previous studies within a reasonable limit. The major reason behind the variation is that this study only considers the copper beneficiation process, whereas most of the previous studies considered the entire mining process.

Another key point to be noted is economy-based allocation technique is considered in this study, whereas most of the previous studies considered mass-based allocation. The overall impact of zinc in GHG values in per tonne of zinc production reported in the previous studies agrees very well with the current research. For gold mining, previous studies reported that GHG varies with 180 t $CO_2$ eq/kg Au, which was based on hybrid allocation based on both mass and economy. This research only considered economy-based allocation, which yields more turnover of environmental impacts on gold beneficiation due to its high value. For silver and lead, no significant research output is reported in the open literature based on the environmental impacts through LCA.

Environmentally sustainable beneficiation processes are currently in practice or under development not only for gold—silver—lead—zinc—copper but also for other metals. This sustainable production processes can reduce chemical emissions and can promote greater sustainability to the environment. Birich et al. presented an alternative production route for silver extraction, which will separate around 60% of nonsilver minerals, and afterwards, nitric acid leaching will produce silver with concentrations up to 98%. The silver is separated through copper cementation (Birich et al., 2018). Andrea et al. presented a novel approach for rare-earth metal separation in Europe through separation based on eudialyte (Schreiber et al., 2016). Azimi et al. presented a dry coal beneficiation method, which would reduce environmental burdens (Rajender Gupta, 2014). Edy et al. presented bio flotation techniques for iron and sulphur, which is based on bacteria and mineral interaction for sustainable production of minerals (Sanwani et al., 2016). Ergin et al. presented environment-friendly optical separation of minerals like lignite, which could reduce the environmental impacts (Gülcan and Gülsoy, 2017). Natalya et al. presented a two-step hydrometallurgical technique to reduce environmental emissions from copper—zinc concentrate processing (Fomchenko and Muravyov, 2018). Natarajan et al. studied biotechnology, microbial- and bioleaching-based beneficiation process of metals (Natarajan, 2006). A similar study has been conducted by Silva et al. and Sukla et al. for metal extraction and beneficiation (Behari et al., 2014; Silva et al., 2017). Lijun et al. presented pumped storage based coal beneficiation, which would increase the energy efficiency of coal mining (Zhang et al., 2014). There are more studies which are working on upgrading the flotation technique of metals and minerals, which would result in sustainable production practices (Feng et al., 2018a, 2018c, 2017c).

## Limitations and Future Recommendations

According to the findings of this study, the major source of the environmental impact of metal beneficiation process is energy consumption in the form of fossil fuel for electricity generation and the blasting process. Carbon-14 emission from medium voltage electricity generation is responsible for ionising radiation. On the other hand, blasting releases nitrogen oxides and zinc; this causes photochemical ozone formation, terrestrial eutrophication, marine eutrophication, human toxicity (noncancer effects) and acidification. Replacing fossil fuels with renewable energy resources would be a great solution to reduce these environmental burdens. A sensitivity analysis presented in this chapter showed that different electricity grid mix with more nonfossil fuels could significantly reduce the environmental effects of gold beneficiation. Energy integration can be done by replacing electricity generation sources or replacing the process of heat generation resources. Solar industrial process heating systems are already in operation for mining industries in Chile, South Africa and Oman (Farjana et al., 2018d). Many other research works are already in progress to assess the prospect of energy efficiency and energy integration in mining industries (Eglinton et al., 2013; Günter and Colin, 2016; Paraskevas et al., 2016). Though energy integration would be a great sustainability solution for the mining sector, it implies the demerits of significant capital cost and the availability of renewable energy during off-peak hours (Farjana et al., 2018d, 2018e). To avail the renewable energy sources, additional energy storage systems could be utilised that can also impose a larger capital cost.

In summary, the recommendation of this chapter is to integrate renewable energy generation systems into the metal beneficiation and flotation processes. Further assessment is required to analyse the feasibility of energy integration, process integration and life cycle cost analysis (CAPEX and OPEX). This study provides invaluable information on the fact that metal beneficiation process is not only affecting global climate change but also is harmful to human health through ionising radiation effect to a great extent. However, the results reported in this study are based on only the economy-based allocation method, which implies more weights to the environmental emission from gold beneficiation.

## Conclusion

In conclusion, this chapter presents and discusses the key factors that cause significant environmental impacts due to gold—silver—lead—zinc—copper beneficiation process. The major environmental impact categories are analysed using the ILCD method and the CED method. The primary reasons behind the environmental burdens associated during the beneficiation process are the blasting process and the amount of electricity consumption in the beneficiation process. From the analysis results presented here, the gold and silver metal beneficiation processes are the most impactful towards environmental sustainability. Gold beneficiation is dominant over the other metals, whereas ionising radiation has the highest impact. The improvement of energy generation source grades and improvement in energy efficiency would be most helpful to reduce the environmental effects on human health, ecosystem and biodiversity. Modification of the grade of the fuel resources or improvement of energy efficiency would be helpful to turn the beneficiation technologies into an environmentally friendly industrial processing route.

## References

Althaus, H.J., Classen, M., 2005. Life cycle inventories of metals and methodological aspects of inventorying material resources in ecoinvent. Int. J. Life Cycle Assess. 10, 43–49. https://doi.org/10.1065/lca2004.11.181.5.

Ashraf, M.A., Sarfraz, M., Naureen, R., Gharibreza, M., 2015. Environmental Impacts of Metallic Elements. https://doi.org/10.1007/978-981-287-293-7.

Awuah-Offei, K., Adekpedjou, A., 2011. Application of life cycle assessment in the mining industry. Int. J. Life Cycle Assess. 16, 82–89. https://doi.org/10.1007/s11367-010-0246-6.

Behari, L., Council, S., Researc, I., Technol, M., Technol, M., Technol, M., 2014. Biomineral processing. A valid eco-friend. alter. metal extract. In: https://www.researchgate.net/publication/266910385_Biomineral_Processing_A_Valid_Eco-Friendly_Alternative_for_Metal_Extraction.

Birich, A., Friedrich, B., Gronen, L., Katzmarzyk, J., Silin, I., Wotruba, H., 2018. Alternative silver production by environmental sound processing of a sulfo salt silver mineral found in Bolivia. Metals 8. https://doi.org/10.3390/met8020114.

Canda, L., Heput, T., Ardelean, E., 2016. Methods for recovering precious metals from industrial waste. IOP Conf. Ser. Mater. Sci. Eng. 106 https://doi.org/10.1088/1757-899X/106/1/012020.

Chen, W., Geng, Y., Hong, J., Dong, H., Cui, X., Sun, M., Zhang, Q., 2018. Life cycle assessment of gold production in China. J. Clean. Prod. 179, 143–150. https://doi.org/10.1016/j.jclepro.2018.01.114.

Christie, T., Brathwaite, B., 1995. Mineral Commodity Report 6 — Lead and Zinc, vol. 16. New Zeal. Min, pp. 22–30.

Corti, C.W., Holliday, R.J., 2004. Commercial aspects of gold applications: from materials science to chemical science. Gold Bull. 37, 20–26. https://doi.org/10.1007/BF03215513.

Davis, J.R., 2001. Aluminum and aluminum alloys. Light Met. Alloy 66. https://doi.org/10.1361/autb2001p351.

Drzymala, J., 2007. Mineral Processing, Foundations of Theory and Practice of Minerallurgy. https://doi.org/10.1017/CBO9781107415324.004.

Eglinton, T., Hinkley, J., Beath, A., Dell'Amico, M., 2013. Potential applications of concentrated solar thermal technologies in the Australian minerals processing and extractive metallurgical industry. J. Occup. Med. 65, 1710–1720. https://doi.org/10.1007/s11837-013-0707-z.

Farjana, S.H., Huda, N., Mahmud, M.A.P., 2018. Environmental Impact Assessment of European Non-ferro Mining Industries through Life-Cycle Assessment Environmental Impact Assessment of European Non-ferro Mining Industries through Life-Cycle Assessment 0–7.

Farjana, S.H., Huda, N., Mahmud, M.A.P., Lang, C., 2018. Comparative life-cycle assessment of uranium extraction processes in Australia. J. Clean. Prod. 202, 666−683. https://doi.org/10.1016/j.jclepro.2018.08.105.

Farjana, S.H., Huda, N., Mahmud, M.A.P., Lang, C., 2018. Life-Cycle Environmental Impact Assessment of Mineral Industries Life-Cycle Environmental Impact Assessment of Mineral Industries. https://doi.org/10.1088/1757-899X/351/1/012016.

Farjana, S.H., Huda, N., Mahmud, M.A.P., Saidur, R., 2018d. Solar process heat in industrial systems − a global review. Renew. Sustain. Energy Rev. 82, 2270−2286. https://doi.org/10.1016/j.rser.2017.08.065.

Farjana, S.H., Huda, N., Mahmud, M.A.P., Saidur, R., 2018. Solar industrial process heating systems in operation − current SHIP plants and future prospects in Australia. Renew. Sustain. Energy Rev. 91 https://doi.org/10.1016/j.rser.2018.03.105.

Fashola, M.O., Ngole-Jeme, V.M., Babalola, O.O., 2016. Heavy metal pollution from gold mines: environmental effects and bacterial strategies for resistance. Int. J. Environ. Res. Publ. Health 13. https://doi.org/10.3390/ijerph13111047.

Feng, Q., Wen, S., 2017. Formation of zinc sulfide species on smithsonite surfaces and its response to flotation performance. J. Alloys Compd. 709, 602−608. https://doi.org/10.1016/j.jallcom.2017.03.195.

Feng, Q., Wen, S., Deng, J., Zhao, W., 2017. Combined DFT and XPS investigation of enhanced adsorption of sulfide species onto cerussite by surface modification with chloride. Appl. Surf. Sci. 425, 8−15. https://doi.org/10.1016/j.apsusc.2017.07.017.

Feng, Q., Wen, S., Zhao, W., Chen, H., 2018. Interaction mechanism of magnesium ions with cassiterite and quartz surfaces and its response to flotation separation. Separ. Purif. Technol. 206, 239−246. https://doi.org/10.1016/j.seppur.2018.06.005.

Feng, Q., Zhao, W., Wen, S., 2018. Surface modification of malachite with ethanediamine and its effect on sulfidization flotation. Appl. Surf. Sci. 436, 823−831. https://doi.org/10.1016/j.apsusc.2017.12.113.

Feng, Q., Zhao, W., Wen, S., 2018. Ammonia modification for enhancing adsorption of sulfide species onto malachite surfaces and implications for flotation. J. Alloys Compd. 744, 301−309. https://doi.org/10.1016/j.jallcom.2018.02.056.

Feng, Q., Zhao, W., Wen, S., Cao, Q., 2017. Activation mechanism of lead ions in cassiterite flotation with salicylhydroxamic acid as collector. Separ. Purif. Technol. 178, 193−199. https://doi.org/10.1016/j.seppur.2017.01.053.

Feng, Q., Zhao, W., Wen, S., Cao, Q., 2017. Copper sulfide species formed on malachite surfaces in relation to flotation. J. Ind. Eng. Chem. 48, 125−132. https://doi.org/10.1016/j.jiec.2016.12.029.

Fomchenko, N.V., Muravyov, M.I., 2018. Two-step biohydrometallurgical technology of copper-zinc concentrate processing as an opportunity to reduce negative impacts on the environment. J. Environ. Manag. 226, 270−277. https://doi.org/10.1016/j.jenvman.2018.08.045.

Goedkoop, M., Oele, M., Vieira, M., Leijting, J., Ponsioen, T., Meijer, E., 2014. SimaPro Tutorial, vol. 89. https://doi.org/10.1142/S0218625X03005293.

Gülcan, E., Gülsoy, Ö.Y., 2017. Performance evaluation of optical sorting in mineral processing − a case study with quartz, magnesite, hematite, lignite, copper and gold ores. Int. J. Miner. Process. 169, 129−141. https://doi.org/10.1016/j.minpro.2017.11.007.

Günter, R., Colin, A., 2016. A literature review on the potential of renewable electricity sources for mining operations in South Africa. J. Energy South Afr. 27, 1−21.

Haque, N., Norgate, T., 2014. The greenhouse gas footprint of in-situ leaching of uranium, gold and copper in Australia. J. Clean. Prod. 84, 382−390. https://doi.org/10.1016/j.jclepro.2013.09.033.

Lee, P., Kwon, J., Lee, J., Lee, H., Suh, Y.D., Hong, S., Yeo, J., 2017. Rapid and effective electrical conductivity improvement of the Ag NW-based conductor by using the laser-induced nano-welding process. Micromachines 8, 1−10. https://doi.org/10.3390/mi8050164.

Lima, F.M., Lovon-Canchumani, G.A., Sampaio, M., Tarazona-Alvarado, L.M., 2018. Life cycle assessment of the production of rare earth oxides from a Brazilian ore. Procedia CIRP 69, 481−486. https://doi.org/10.1016/j.procir.2017.11.066.

Long, K.R., DeYoung, J.H., Ludington, S.D., 1998. Database of Significant Deposits of Gold, Silver, Copper, Lead, and Zinc in the United States. Part A: Database Description and Analysis. Open-File Report 98-206A. US Geol. Surv., pp. 1−60

Mahmud, M., Huda, N., Farjana, S., Lang, C., Mahmud, M.A.P., Huda, N., Farjana, S.H., Lang, C., 2018. Environmental impacts of solar-photovoltaic and solar-thermal systems with life-cycle assessment. Energies 2018 11. https://doi.org/10.3390/EN11092346, 2346 11, 2346.

Mahmud, M.A.P., Huda, N., Farjana, S.H., 2018. Environmental Profile Evaluations of Piezoelectric Polymers Using Life Cycle Assessment Environmental Profile Evaluations of Piezoelectric Polymers Using Life Cycle Assessment.

Mahmud, M.A.P., Huda, N., Farjana, S.H., Lang, C., 2018. Environmental sustainability assessment of hydropower plant in Europe using life cycle assessment. In: IOP Conference Series: Materials Science and Engineering. https://doi.org/10.1088/1757-899X/351/1/012006.

Mahmud, M.A.P., Huda, N., Farjana, S.H., Lang, C., 2018. Environmental life-cycle assessment and techno-economic analysis of photovoltaic ( PV ) and photovoltaic/thermal ( PV/T ) systems. 2018. In: IEEE Int. Conf. Environ. Electr. Eng. 2018 IEEE Ind. Commer. Power Syst. Eur. (EEEIC/I&CPS Eur., pp. 1−5.

Memary, R., Giurco, D., Mudd, G., Mason, L., 2012. Life cycle assessment: a time-series analysis of copper. J. Clean. Prod. 33, 97−108. https://doi.org/10.1016/j.jclepro.2012.04.025.

Natarajan, K.A., 2006. Biotechnology for metal extraction, mineral beneficiation and environmental control. Proc. Int. Semin. Miner. Process. Technol. 68−81.

Norgate, T., Haque, N., 2012. Using life cycle assessment to evaluate some environmental impacts of gold production. J. Clean. Prod. 29−30, 53−63. https://doi.org/10.1016/j.jclepro.2012.01.042.

Norgate, T.E., 2001. A Comparative Life Cycle Assessment of Copper Production Processes.

Norgate, T.E., Jahanshahi, S., Rankin, W.J., 2007. Assessing the environmental impact of metal production processes. J. Clean. Prod. 15, 838−848. https://doi.org/10.1016/j.jclepro.2006.06.018.

Norgate, T.E., Rankin, W.J., 2000. Life Cycle Assessment of copper and nickel production. In: Int. Conf. Miner. Process. Extr. Metall. (MINPREX), Melbourne, 11 - 14 Sept. 2000, pp. 133−138.

Northey, S., Haque, N., Mudd, G., 2013. Using sustainability reporting to assess the environmental footprint of copper mining. J. Clean. Prod. 40, 118−128. https://doi.org/10.1016/j.jclepro.2012.09.027.

Northey, S.A., Haque, N., Lovel, R., Cooksey, M.A., 2014. Evaluating the application of water footprint methods to primary metal production systems. Miner. Eng. 69, 65−80. https://doi.org/10.1016/j.mineng.2014.07.006.

Paraskevas, D., Kellens, K., Van De Voorde, A., Dewulf, W., Duflou, J.R., 2016. Environmental impact analysis of primary aluminium production at country level. Procedia CIRP 40, 209−213. https://doi.org/10.1016/j.procir.2016.01.104.

Qi, C., Ye, L., Ma, X., Yang, D., Hong, J., 2017. Life cycle assessment of the hydrometallurgical zinc production chain in China. J. Clean. Prod. 156, 451−458. https://doi.org/10.1016/j.jclepro.2017.04.084.

Rajender Gupta, E.A., 2014. Dry coal beneficiation method-effective key to reduce health, environmental and technical coal firing issues. J. Civ. Environ. Eng. 04 https://doi.org/10.4172/2165-784X.1000e118.

Sanwani, E., Chaerun, S., Mirahati, R., Wahyuningsih, T., 2016. Bioflotation: bacteria-mineral interaction for eco-friendly and sustainable mineral processing. Procedia Chem. 19, 666−672. https://doi.org/10.1016/j.proche.2016.03.068.

Schreiber, A., Marx, J., Zapp, P., Hake, J.-F., Voßenkaul, D., Friedrich, B., 2016. Environmental impacts of rare earth mining and separation based on eudialyte. A New Euro. Way. Res. 5, 32. https://doi.org/10.3390/resources5040032.

Shedd, K.B., 2016. Mineral Commodity Summaries. Usgs 0, 2.

Silva, R.A., Borja, D., Hwang, G., Hong, G., Gupta, V., Bradford, S.A., Zhang, Y., Kim, H., 2017. Analysis of the effects of natural organic matter in zinc beneficiation. J. Clean. Prod. 168, 814−822. https://doi.org/10.1016/j.jclepro.2017.09.011.

Van Genderen, E., Wildnauer, M., Santero, N., Sidi, N., 2016. A global life cycle assessment for primary zinc production. Int. J. Life Cycle Assess. 21, 1580−1593. https://doi.org/10.1007/s11367-016-1131-8.

Wuana, R.A., Okieimen, F.E., 2011. Heavy metals in contaminated soils: a review of sources, chemistry, risks and best available strategies for remediation. ISRN Ecol 2011, 1−20. https://doi.org/10.5402/2011/402647.

Zhang, L., Wang, J., Feng, Y., 2018. Life cycle assessment of opencast coal mine production: a case study in Yimin mining area in China. Environ. Sci. Pollut. Res. 25, 8475−8486. https://doi.org/10.1007/s11356-017-1169-6.

Zhang, L., Xia, X., Zhang, J., 2014. Improving energy efficiency of cyclone circuits in coal beneficiation plants by pump-storage systems. Appl. Energy 119, 306−313. https://doi.org/10.1016/j.apenergy.2014.01.031.

Zuazo, V.H.D., Pleguezuelo, C.R.R., 2008. Soil-erosion and runoff prevention by plant covers . A review to cite this version : HAL Id : hal-00886458 Soil-erosion and runo ff prevention by plant covers. A Rev. Agron. Sustain. Dev. 28 (28), 65−86. https://doi.org/10.1051/agro:2007062.

Chapter 6

# Life Cycle Assessment of Solar Process Heating System Integrated in Mining Process

## Introduction

Fossil fuels like coal, natural gas and diesel are widely used by the mining industries to generate electricity, process heat and drive different types of machinery. Usage of these fuels emits large quantities of pollutants like nitrous oxides, carbon dioxides, carbon monoxides, sulphur dioxide and suspended solids like particulate matters (Beath, 2012; Norgate and Haque, 2010). To reduce the environmental impacts associated with the usage of fossil fuel in the mining industries, it is apparent that integrating renewable energy resources for power or heat generation would bring significant benefits According to IRENA, more than 80% of the industrial process heat should be supplied by renewables by 2030 to reach the sustainable development goal (IEA-ETSAP, IRENA, 2015; International Renewable Energy Agency, 2014). A significant share of this energy consumption is in the form of thermal energy. Thermal energy is used as industrial process heat to run the mining, extraction, beneficiation or refining operations associated with ferrous and nonferrous metal extraction processes. A large share of industrial process heat is in low-temperature processes which can be replaced merely by solar energy integration. There are also medium-temperature and high-temperature industrial process heat applications which have the potential to integrate renewables. Solar energy integration in the mining industries requires significant effort to assess the comparative analysis of how these can reduce environmental burdens with the optimised design of the solar network. A typical solar heating system consists of a solar collector, pump, heat exchanger and a storage tank. Solar collectors are the most important component of the solar heating system. They can be of several types, depending on their structures and requirements (Hess et al., 2011; Lillo et al., 2017; Sharma et al., 2017). For example, the flat plate solar collectors are dependable, durable, retain high energy efficiency and supplies low-temperature process heat below 80°C. Evacuated tube collectors (ETCs) contain shallow double tubed wall with a vacuum inside, which is also used for

Life Cycle Assessment for Sustainable Mining. https://doi.org/10.1016/B978-0-323-85451-1.00006-8

low-temperature process heat applications but operates with a little higher temperature (80−120°C temperature). Parabolic trough collectors are used for medium- and high-temperature process heating systems. In any solar industrial process heating systems, heat must be adequate to supply the necessary demand, directly available for process heating and economically feasible. The efficiency of these collectors depends on collector type, available solar radiation, tilt angle, temperature level and required output. However, in addition to the economic viability, the environmental advantages are also needed to be assessed to make an informed decision on how solar process heat would be beneficial for sustainability (Duffie et al., 2003; European Solar Thermal Industry Federation, 2006; Norton, 2012). A few mining industries in Chile, South Africa, Cyprus, Austria, India and Germany are utilising solar heat for processes, which includes operations like copper electrowinning, mining and cleaning, nickel and galvanic bath, etc (Farjana et al., 2018a,b,c,d,e,f).

There are several studies in the scientific literature which assessed the environmental loads caused by the mining industries, identified the critical process and reasons of environmental emissions and suggested a possible solution to reduce the environmental burdens. Life cycle assessment (LCA) is a powerful tool utilised by scientists and researchers to assess the ecological loads associated with a process or a system of processes based on standardised methods. LCA-based studies on aluminium production showed that the emission from aluminium production is mostly due to alumina smelting and refining operations, which consume a huge amount of electricity. Process heat consumption and blasting operations are also responsible for environmental impacts (Farjana et al., 2019; Nunez and Jones, 2016; Tan and Khoo, 2005). LCA studies on coal mining assessed the sensitivity of the mining processes due to the mining tailing management processes, fugitive emission factor, consumption of diesel oil and the emission factor for diesel oil combustion (Adiansyah et al., 2017; Guimarães da Silva et al., 2018). LCA studies based on copper mining showed that solvent extraction and electrowinning processes are the largest emitters of pollutants. However, the diesel consumptions for process heat generation also possess significant environmental burdens associated with mining processes (Memary et al., 2012; Norgate and Rankin, 2000; Northey et al., 2013). Farjana et al. analysed the LCA of ilmenite and rutile production process, which impose environmental burdens due to the mining process. Mining process consumes a large number of fossil fuels to generate electricity and process heat (Farjana et al., 2018c,d,e,f). LCA research on gold mining investigated the electricity generation, sulphur concentrate production and gold ore mining processes (Chen et al., 2019, 2018). Other studies on the environmental burden associated with gold beneficiation and refining operation were assessed (Farjana et al., 2018c,d,e,f; Farjana et al., 2019a,c). For magnesium oxide, a sensitivity analysis was conducted based on different energy mix scenario (Ruan and Unluer, 2016). LCA studies on nickel mining show that impacts are due to fossil fuel consumption in nickel smelting and refining and ferronickel consumption in high-pressure acid leaching (Bartzas and Komnitsas, 2015; Khoo et al., 2017; Norgate and Rankin, 2000). LCA studies based on uranium mining show that fuel enrichment stage made the most significant environmental loads, along with the mining process, due to the vast amount of electricity and process heat consumption (Farjana et al., 2018c,d,e,f; Haque and Norgate, 2014).

Several LCA studies focussed on the environmental impact analysis in mining industries that were focussed on identifying scientific ways to reduce environmental burdens associated with mining processes or subprocesses (Farjana et al., 2018a,b,c,e; Farjana et al., 2019b,d). The processes may vary depending on the specific use of technology mix (technology of the mining processes), energy mix (based on different forms of energy) and electricity mix (based on the electricity generation sources, electricity mix or grid mix scenario). Only a few studies investigated using the different energy mix scenario for mining industries. However, none of them assessed the energy mix scenario for process heat generation sources utilised to provide thermal energy involved in the mining processes. This novel contribution conducted extensive research based on the energy mix scenario for the ore processing of seven key mining industries — aluminium, copper, ilmenite, rutile, lead, nickel and uranium. It is indispensable to conduct a systematic analysis to assess the possible emission reduction potential of solar industrial process heating systems in mining processes, which is the novel contribution of this chapter. This is by far the first ever reported work in the open literature to conduct such an extensive study focussed on seven important mining and mineral processing industries. The study analysed the base case of the current scenario of mining and extraction processes of seven major mining industries, followed by integrated solar process heating analysis based on two types of solar collector. The first one considered the integration of flat plate collector (FPC) to supply the process heat, and the second case considered the integration of ETCs to supply the process heat. The results of these three cases are compared to evaluate the emission reduction potentials of these mining processes using solar process heat.

## Case Study of Life Cycle Assessment

LCA is an internationally recognised environmental impact assessment tool to analyse the environmental effects of a product, process or a system of operations on the environment. The LCA methodology is built on the ISO standard methodology ISO 14040. According to these ISO standards, the LCA can be subdivided into four significant steps: goal and scope definition, life cycle inventory (LCI), LCA and results in interpretation (Fogler and Timmons, 1998; Mahmud et al., 2018a,b, 2019). In this chapter, the goal is to analyse the base case results of seven mining processes, compare among them and carry out sensitivity analysis while replacing the fossil fuels used for process heat generation with solar process heat. Three scenarios are then compared among these mining processes. The scope of this LCA study includes environmental emissions which are emitted from these mining operations. In the LCI, the fossil fuels used for process heat and electricity generation, electricity, materials, inorganic and organic chemicals, water and transportation are considered as material inputs. The produced metal, heat emissions, chemical pollutant emission and waste in other forms are regarded as material outputs. The functional unit is chosen as 1 kg of ore mined metal. The system boundary for each of the metals considered only includes the ore mining step. The system boundary followed for each metal is cradle-to-gate.

The analysis is done using SimaPro software version 8.5. The datasets are originated from several sources which are gathered and documented in the EcoInvent database (Goedkoop et al., 2014; PRé, 2018). The LCI data for bauxite mining are originated from the datasets provided by the International Aluminium Institute (Nunez and Jones, 2016). The LCI data for copper ore mining, lead ore mining and nickel ore mining are gathered from the literature from Classen et al. (Althaus and Classen, 2005). The LCI data for ilmenite and rutile ore mining are collected from Althaus et al. (2007). Finally, the LCI dataset for uranium ore mining is originated from Doka ISL datasets (Doka, 2011). The geographic coverage for these metals is mostly global, except for ilmenite and rutile ore mining. Global datasets are prioritised based on their availability as it is hard to match and collect location-specific datasets. As the scope of this chapter only involves the ore mining stage for each metal, it is assumed that no metal coproducts are going outside the system unless the 1 kg of ore is mined and ready for beneficiation process. The LCA methodology employed here is the IMPACT 2002+ method, which is both a midpoint indicator and damage-oriented approach. The methodology is adopted from EcoIndicator 99 method, CML method, IPCC 2001 method and other methods from literature documented in EcoInvent. The midpoint indicator-based categories of IMPACT 2002+ are aquatic ecotoxicity, terrestrial ecotoxicity, terrestrial acidification, aquatic acidification, aquatic eutrophication, carcinogens, noncarcinogens, respiratory organics, respiratory inorganics, ionising radiation, global warming, land occupation, renewable energy and mineral extraction. The damage categories are human health, ecosystem quality, resources and climate change (Hischier et al., 2010; Humbert et al., 2012; Lehtinen et al., 2011). This study is conducted based on three different scenarios for each of the metals — aluminium, copper, ilmenite, rutile, lead, nickel and uranium. The section below describes those cases in detail. Case 1 represents the base case scenario without energy integration; case 2 describes the flat plate solar collector integrated ore mining processes. Case 3 describes the evacuated tube solar collector integrated ore mining processes. Table 42 describes the LCI inputs to produce 1 kg of mined ore.

**TABLE 42** Life cycle inventory datasets for seven metals.

| Material input | Unit | Aluminium (EPA, 2018) | Copper (Althaus and Classen, 2005) | Ilmenite (Althaus et al., 2007) | Rutile (Althaus et al., 2007) | Nickel (Althaus and Classen, 2005) | Lead (Althaus and Classen, 2005) | Uranium (Doka, 2011) |
|---|---|---|---|---|---|---|---|---|
| Aluminium hydroxide | P | | 4.92E-10 | | | 6.71E-10 | | |
| Ammonia | kg | | | | | 0.08 | | 0.9 |
| Ammonium sulphate | kg | | | | | | | 0.106 |
| Blasting | kg | | 0.08 | | | 0.12 | | |
| Carbon monoxide, CO | kg | | | | | | | |
| Chemicals, organic | kg | | 0.013 | | | 0.018 | | 0.315 |
| Chemicals, inorganic | kg | | 0.045 | | | 0.06 | | 0.26 |
| Conveyor belt | m | | 2.24E-06 | | | 3.05E-06 | | |
| Copper, 1.13% in sulphide, Cu 0.76% and Ni 0.76% in crude ore | kg | | 1.26 | | | | | |
| Diesel, combusted in industrial boiler | l | 4.30E-03 | | | | | | |
| Diesel burned in the diesel-electric generating set | MJ | | | | | | | 176 |

*Continued*

**TABLE 42** Life cycle inventory datasets for seven metals.—cont'd

| Material input | Unit | Aluminium (EPA, 2018) | Copper (Althaus and Classen, 2005) | Ilmenite (Althaus et al., 2007) | Rutile (Althaus et al., 2007) | Nickel (Althaus and Classen, 2005) | Lead (Althaus and Classen, 2005) | Uranium (Doka, 2011) |
|---|---|---|---|---|---|---|---|---|
| Diesel burned in building machine | MJ | | 5.96 | 0.286 | 1.48 | 8.1 | | |
| Electricity, medium voltage | kWh | 3.90E-04 | 0.92 | 0.16 | 0.87 | 1.26 | 0.15 | |
| Electricity, high voltage | kWh | | 0.73 | | | 3.73 | | |
| Electricity, hydropower | kWh | | 3.40E-03 | | | 8.77 | | |
| Ethylenediamine | kg | | | | | | | 0.012 |
| Gasoline | l | 2.60E-04 | | | | | | |
| Heat, natural gas, at industrial furnace >100 kW | MJ | | 1.78 | 0.198 | 1.03 | 14.47 | 0.55 | |
| Heat, at hard coal industrial furnace 1–10 MW | MJ | | 0.64 | 0.66 | 3.42 | 1.86 | 6.74 | |
| Heavy fuel oil, burned in industrial furnace 1 MW | MJ | | 10.14 | 0.066 | 0.34 | 22.67 | 0.27 | 264 |

| Metric | Unit | | | | | | | |
|---|---|---|---|---|---|---|---|---|
| Hydrogen cyanide | kg | | | 2.70E-03 | | | 2.05E-03 | |
| Hydrogen, liquid | kg | | | 4.50E-03 | | | | |
| Iron ore, 65% Fe | Kg | | 0.06 | | | | | |
| Lead concentrate | Kg | | 1.79 | | | | | |
| Limestone, milled, packed | Kg | | 0.64 | 1.92 | | | 0.62 | |
| Mine, bauxite | P | | | | 1.14E-09 | 2.2E-10 | | |
| Nitrogen | Kg | | 0.01 | | | | | |
| Nickel, 1.13% in sulphide, Ni 0.76% and Cu 0.76% in crude ore, in-ground | Kg | | | 1.26 | | | | |
| Nonferrous metal mine | P | | | 0.08 | | | 2.93E-09 | |
| Nonferrous metal smelter | P | | | 3.35E-11 | | | 2.46E-11 | |
| Occupation, the mineral extraction site | $m^2a$ | | | | 0.07 | 0.014 | | |
| Occupation, dump site | $m^2a$ | 1.96 | | | | | | |
| Oxygen | kg | | 0.2 | | | | 1.92 | |
| Portland calcareous cement | kg | | | 2.62 | | | | |
| Recultivation, bauxite mine | $m^2$ | | | | 7.70E-03 | 1.40E-03 | | |
| Residual fuel oil | l | | | | | | | 1.20E-03 |

*Continued*

**TABLE 42** Life cycle inventory datasets for seven metals.—cont'd

| Material input | Unit | Aluminium (EPA, 2018) | Copper (Althaus and Classen, 2005) | Ilmenite (Althaus et al., 2007) | Rutile (Althaus et al., 2007) | Nickel (Althaus and Classen, 2005) | Lead (Althaus and Classen, 2005) | Uranium (Doka, 2011) |
|---|---|---|---|---|---|---|---|---|
| Resource correction, PbZn, silver, negative | kg | | | | | | 1.50E-03 | |
| Resource correction, PbZn, lead, positive | kg | | | | | | 0.035 | |
| Sand | kg | | 24.28 | | | 33.09 | | |
| Silica sand | kg | | 0.6 | | | 1.89 | 0.6 | |
| Soda, powder | kg | | | | | | | 2.5 |
| Sodium chlorate, powder | kg | | | | | | | 1 |
| Sodium chloride, brine solution | kg | | | | | | | 2.5 |
| Sodium hydroxide | kg | | | | | | | 0.026 |
| TiO$_2$, 54% in ilmenite, 2.6% in crude ore | kg | | | 0.56 | | | | |
| TiO$_2$, 95% in rutile, 0.40% in crude ore, in ground | kg | | | | 1 | | | |

| | | | | | | | | |
|---|---|---|---|---|---|---|---|---|
| Transport, ocean freighter, residual fuel oil | tkm | 2.5 | | | | | | |
| Transport, ocean freighter, diesel | tkm | 0.28 | | | | | 5.27 | 32 |
| Transport, lorry >16t, the fleet average | | | 1.32 | | | 1.79 | 0.93 | 6.3 |
| Transformation, from the forest | m² | | | 5.90E-04 | 3.08E-03 | | | |
| Transformation, from pasture and meadow | m² | | | 8.90E-04 | 4.60E-03 | | | |
| Transformation, to the mineral extraction site | m² | | | 1.40E-03 | 7.70E-03 | | | |
| Transformation, from unknown | m² | | | | | | | 0.018 |
| Transformation, to dump site | m² | | | | | | | 0.018 |
| Uranium mill | p | | | | | | | 1.35E-07 |
| Uranium natural | kg | | | | | | | 1.05 |
| Water, river | m³ | | 0.02 | | | 0.027 | | |
| Water, well | m³ | | 0.12 | 9.60E-03 | 0.05 | 0.16 | | |
| Water, unspecified natural origin | m³ | | | | | | | 1 |

## Case 1: Base Case

The base case scenario contains the actual measures and type of fossil fuels used for process heat generation. Mostly the diesel oil, natural gas and residual fuel oil are used for process heat generation. According to the inventory datasets presented in Table 43, for bauxite ore mining, the process heat is generated using residual fuel oil, gasoline and diesel. For copper ore mining, ilmenite and rutile ore mining, the process heat is generated using hard coal, natural gas and heavy fuel oil. For uranium ore mining, process heat is supplied after combustion of heavy fuel oil. The quantity of the fossil fuel used for process heat generation varies from one metal mining process to another. For example, in the considered methods the diesel is used in bauxite ore mining here which is 4.3E-03 L, residual fuel oil is 1.2E-03 L and gasoline is 2.6E-04 L. The heat generated using natural gas is 1.78 MJ for copper ore mining, 0.198 MJ for ilmenite mining, 1.03 MJ for rutile mining, 14.47 MJ for nickel mining and 0.55 MJ for lead ore mining. The heat generated using hard coal is 0.64 MJ for copper ore mining, 0.66 MJ for ilmenite mining, 3.42 MJ for rutile mining, 1.86 MJ for nickel mining and 6.74 MJ for lead ore mining. The heat generated using heavy fuel oil is 10.14 MJ for copper ore mining, 0.066 MJ for ilmenite mining, 0.34 MJ for rutile mining, 22.67 MJ for nickel mining, 0.27 MJ for lead ore mining and 264 MJ for uranium ore mining.

## Case 2: Flat Plate Collector-based System

The second case considers seven ore mining systems integrated with flat plate solar collector technologies to supply process heat. In the flat plate solar collector system to produce 1 MJ process heat, amount of electricity used is 0.009 kWh; the solar system power is 2.19E-06 p, transport 0.000241 tkm and heat waste 1.158 MJ. No fossil fuel is used for process heat generation. In case of bauxite ore mining, amount of total process heat is 5.76E-08 MJ, for copper ore mining it is 12.56 MJ, for ilmenite ore mining it is 0.924 MJ, for rutile ore mining it is 4.79 MJ, for nickel ore mining it is 39 MJ, for lead ore mining it is 7.56 MJ and for uranium ore mining it is 264 MJ.

## Case 3: Evacuated Tube Collector-based System

In the third case, seven ore mining systems are integrated with evacuated tube solar collector technologies to supply process heat. In the evacuated tube solar collector system to produce 1 MJ process heat, amount of electricity used is 0.00957 kWh; the solar system power is 2.12E-06 p, transport 0.000235 tkm and heat waste 1.167 MJ. Like case 2, no fossil fuel is used for process heat generation. The amount of fossil fuel used for process heat generation replaced by solar heat is the same in quantity like case 2. In case of bauxite ore mining, amount of total process heat is 5.76E-08 MJ, for copper ore mining it is

**TABLE 43** Life cycle assessment results for three scenarios on the categories of human health.

| Impact category | Unit | Carcinogens kg C₂H₃Cl eq | Noncarcinogens kg C₂H₃Cl eq | Respiratory inorganics kg PM2.5 eq | Respiratory organics kg C₂H₄ eq | Ionising radiation Bq C-14 eq |
|---|---|---|---|---|---|---|
| Aluminium | Base case | 4.47E-05 | 5.20E-03 | 2.10E-04 | 4.70E-05 | 0 |
|  | FPC | 2.05E-05 | 3.80E-03 | 2.00E-04 | 3.90E-05 | 8.90E-05 |
|  | ETC | 2.02E-05 | 3.80E-03 | 2.00E-04 | 3.90E-05 | 1.40E-04 |
| Copper | Base case | 0.033 | 0.09 | 0.042 | 4.80E-03 | 32.15 |
|  | FPC | 0.029 | 0.09 | 0.04 | 4.60E-03 | 32.34 |
|  | ETC | 0.029 | 0.09 | 0.04 | 4.60E-03 | 32.45 |
| Ilmenite | Base case | 5.2E-04 | 1.1E-03 | 2.8E-04 | 6.31E-05 | 6E-03 |
|  | FPC | 6.4E-04 | 1.1E-03 | 2.2E-04 | 6E-05 | 0.019 |
|  | ETC | 5.8E-04 | 1.06E-03 | 2.2E-04 | 5.98E-05 | 0.027 |
| Rutile | Base case | 2.70E-03 | 5.70E-03 | 1.40E-03 | 3.20E-04 | 0.031 |
|  | FPC | 3.30E-03 | 5.70E-03 | 1.10E-03 | 3.10E-04 | 0.099 |
|  | ETC | 3.00E-03 | 5.50E-03 | 1.10E-03 | 3.10E-04 | 0.14 |
| Nickel | Base case | 0.07 | 0.25 | 0.13 | 7.00E-03 | 160.6 |
|  | FPC | 0.06 | 0.25 | 0.13 | 6.70E-03 | 161.25 |
|  | ETC | 0.06 | 0.25 | 0.13 | 6.70E-03 | 161.6 |

Continued

**TABLE 43** Life cycle assessment results for three scenarios on the categories of human health.—cont'd

| Impact category | | Carcinogens | Noncarcinogens | Respiratory inorganics | Respiratory organics | Ionising radiation |
|---|---|---|---|---|---|---|
| | Unit | kg C₂H₃Cl eq | kg C₂H₃Cl eq | kg PM2.5 eq | kg C₂H₄ eq | Bq C-14 eq |
| Lead | Base case | 0.09 | 0.51 | 5.50E-03 | 1.10E-03 | 0.12 |
| | FPC | 0.097 | 0.5 | 5.00E-03 | 1.10E-03 | 0.23 |
| | ETC | 0.096 | 0.5 | 5.00E-03 | 1.10E-03 | 0.3 |
| Uranium | Base case | 7.1 | 68.76 | 1 | 0.125 | 4.09E-06 |
| | FPC | 6.98 | 68.77 | 0.98 | 0.121 | 4.09E-06 |
| | ETC | 6.96 | 68.76 | 0.98 | 0.121 | 4.09E-06 |

12.56 MJ, for ilmenite ore mining it is 0.924 MJ, for rutile ore mining it is 4.79 MJ, for nickel ore mining it is 39 MJ, for lead ore mining it is 7.56 MJ and for uranium ore mining it is 264 MJ. Fig. 27 shows the dimensional sketch and working principle of both FPC and ETC.

## LCA Results: Impact on Human Health

Impacts on human health are characterised by five impact categories, namely carcinogens, noncarcinogens, respiratory inorganics, respiratory organics and ionising radiation. This section presents the LCA results of those five impact categories for three different case scenarios (base case, FPC and ETC) considered in this study. For aluminium and uranium production, impacts are reduced under all the categories except for the ionising radiation in both flat plates- and evacuated tube-based process heating systems. In the impact category of ionising radiation for aluminium production, ETC shows better results as compared to the FTC, while for uranium production, the impact category remains unchanged. For copper ore mining, in both flat plate- and evacuated tube-based process heating systems, impacts are reduced for three impact categories — carcinogens, respiratory inorganics and respiratory organics. The noncarcinogens impact category during copper production remains unchanged by integrating both types of solar process heating systems, while surprisingly in the ionising radiation category, it increases a little bit after integrating FPC and ETC. The reason for this may be attributed to the production processes of FPC and ETC, but needs proper scientific validation before any conclusion can be drawn. For ilmenite and rutile, in both flat plate- and evacuated tube-based process heating systems, impacts are reduced in both

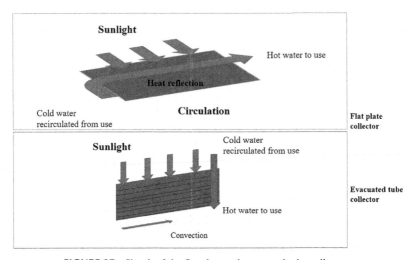

**FIGURE 27**   Sketch of the flat plate and evacuated tube collectors.

categories of respiratory inorganics and respiratory organics. Nickel and lead showed a mixed outcome in all five categories for the three case scenarios considered, as shown in Table 43.

## LCA Results: Impact on Ecosystems Quality

This section describes LCA results on ozone layer depletion, aquatic eco-toxicity, terrestrial ecotoxicity, terrestrial acidification, aquatic acidification and aquatic eutrophication; the detailed results are presented in Table 44. For aluminium mining, solar collector integration would be beneficial to reduce impacts from aquatic ecotoxicity. However, only minor changes are quantified in the other impact categories of terrestrial ecotoxicity, terrestrial acidification, aquatic acidification and aquatic eutrophication. Interestingly, impacts are increased for ozone layer depletion. For copper and nickel ore mining, impacts reduce moderately for ozone layer depletion, aquatic ecotoxicity and terrestrial ecotoxicity. Negligible/no changes in impacts are measured for terrestrial acidification, aquatic acidification and aquatic eutrophication. For ilmenite, rutile, lead and uranium ore mining, impacts reduce for ozone layer depletion, aquatic ecotoxicity and terrestrial ecotoxicity. Only changes appeared from the results of the calculations for terrestrial acidification and aquatic acidification. Between two different types of solar collector, ETC shows slightly better performance for copper, uranium, rutile and uranium ore mining.

## LCA Results: Impact on Climate Change and Resources

This section describes LCA results on climate change, land occupation and nonrenewable energy. Table 45 shows the detailed LCA results from the analysis. For aluminium/bauxite ore mining, global warming impacts and use of nonrenewables decrease due to solar process heat integration. For land occupation, impacts are higher than the base case where ETC shows better performance. For copper, nickel and uranium ore mining, global warming potential is reduced for both solar collectors where ETC shows better results. The impact of land occupation and uses of nonrenewables remain the same for three cases. For ilmenite, rutile and lead ore mining, global warming impacts and land occupation reduce in case of both flat plate and ETC integration and impacts are the same for both integrated solar cases. For the use of non-renewables, results are higher than the base case where ETC is more sustainable.

## LCA Results: Impact based on Damage Categories

This section describes LCA results on damage-based categories results — human health, ecosystems quality and global warming. Table 46 shows the detailed analysis results based on three cases: the base case, the case with FPC

**TABLE 44** Life cycle assessment results for three scenarios on the categories of ecosystems quality.

| Impact category | | Ozone layer depletion | Aquatic ecotoxicity | Terrestrial ecotoxicity | Terrestrial acid | Aquatic acidification | Aquatic eutrophication |
|---|---|---|---|---|---|---|---|
| Unit | | kg CFC-11 eq | kg TEG water | kg TEG soil | kg SO₂ eq | kg SO₂ eq | kg PO₄ P-lim |
| Aluminium | Base case | 2.90E-12 | 34 | 0.014 | 8.50E-03 | 1.30E-03 | 6.26E-07 |
| | FPC | 5.65E-12 | 25.34 | 0.007 | 8.20E-03 | 1.20E-02 | 5.21E-07 |
| | ETC | 5.02E-12 | 25.34 | 0.006 | 8.28E-03 | 1.20E-02 | 5.15E-07 |
| Copper | Base case | 2.97E-07 | 8,205.26 | 2,376.35 | 0.63 | 0.33 | 0.03 |
| | FPC | 1.91E-07 | 8,181.18 | 2,370.017 | 0.62 | 0.32 | 0.03 |
| | ETC | 1.89E-07 | 8,179.16 | 2,369.3 | 0.62 | 0.32 | 0.03 |
| Ilmenite | Base case | 6.52E-09 | 12.29 | 3.63 | 6.6E-03 | 1.3E-03 | 1.23E-05 |
| | FPC | 6.11E-09 | 9.29 | 3.12 | 5.5E-03 | 8.4E-04 | 2E-05 |
| | ETC | 6.015E-09 | 9.143 | 3.075 | 5.5E-03 | 8.3E-04 | 1.9E-05 |
| Rutile | Base case | 3.38E-08 | 63.88 | 18.88 | 0.034 | 6.80E-03 | 6.40E-05 |
| | FPC | 3.17E-08 | 48.26 | 16.27 | 0.029 | 4.30E-03 | 1.00E-04 |
| | ETC | 3.12E-08 | 47.49 | 15.97 | 0.028 | 4.30E-03 | 9.80E-05 |
| Nickel | Base case | 5.73E-07 | 11,834.44 | 3,555.86 | 1.87 | 1.44 | 0.044 |
| | FPC | 3.41E-07 | 11,792.61 | 3,547.81 | 1.85 | 1.43 | 0.044 |
| | ETC | 3.36E-07 | 11,786.33 | 3,545.6 | 1.85 | 1.43 | 0.044 |

*Continued*

**TABLE 44** Life cycle assessment results for three scenarios on the categories of ecosystems quality.—cont'd

| Impact category | | Ozone layer depletion | Aquatic ecotoxicity | Terrestrial ecotoxicity | Terrestrial acid | Aquatic acidification | Aquatic eutrophication |
|---|---|---|---|---|---|---|---|
| Unit | | kg CFC-11 eq | kg TEG water | kg TEG soil | kg SO₂ eq | kg SO₂ eq | kg PO₄ P-lim |
| Lead | Base case | 6.75E-08 | 1,312.07 | 422.78 | 0.12 | 0.047 | 3.50E-03 |
| | FPC | 6.68E-08 | 1,280.2 | 416.93 | 0.1 | 0.043 | 3.60E-03 |
| | ETC | 6.60E-08 | 1,278.98 | 416.5 | 0.1 | 0.043 | 3.60E-03 |
| Uranium | Base case | 9.06E-06 | 1,781,990.4 | 142,546.3 | 5.88 | 1.36 | 0.021 |
| | FPC | 6.26E-06 | 1,781,404.64 | 142,373.7 | 5.62 | 1.219 | 0.022 |
| | ETC | 6.24E-06 | 1,781,362.24 | 142,358.7 | 5.61 | 1.217 | 0.02 |

**TABLE 45** Life cycle assessment results for three scenarios on the categories of global warming and resources.

| | Impact category | Land occupation | Global warming | Nonrenewable energy |
|---|---|---|---|---|
| | Unit | M²org.arable | kg CO₂ eq | MJ primary |
| Aluminium | Base case | 0 | 0.078 | 1.07 |
| | FPC | 4.23E-06 | 0.059 | 0.8 |
| | ETC | 4.10E-06 | 0.059 | 0.8 |
| Copper | Base case | 0.24 | 5.85 | 2.80E-04 |
| | FPC | 0.24 | 4.95 | 2.80E-04 |
| | ETC | 0.24 | 4.93 | 2.80E-04 |
| Ilmenite | Base case | 0.018 | 0.29 | 3.06E-07 |
| | FPC | 0.018 | 0.22 | 4.77E-07 |
| | ETC | 0.018 | 0.22 | 4.68E-07 |
| Rutile | Base case | 0.09 | 1.54 | 1.59E-06 |
| | FPC | 0.095 | 1.18 | 2.47E-06 |
| | ETC | 0.095 | 1.17 | 2.43E-06 |
| Nickel | Base case | 0.37 | 11.84 | 3.90E-04 |
| | FPC | 0.39 | 9.33 | 3.90E-04 |
| | ETC | 0.39 | 9.27 | 3.90E-04 |
| Lead | Base case | 0.07 | 2.34 | 3.26E-05 |
| | FPC | 0.071 | 1.74 | 3.40E-05 |
| | ETC | 0.071 | 1.72 | 3.30E-05 |
| Uranium | Base case | 5.84 | 98.15 | 1.8E0-03 |
| | FPC | 5.95 | 78.66 | 1.8E0-03 |
| | ETC | 5.95 | 78.23 | 1.8E0-03 |

integration and case with ETC integration. According to the results, for the human health category, the highest emission comes from particulates <2.5 mm, nitrogen oxides and sulphur oxides. For bauxite, ilmenite and rutile mining, the highest emission is from nitrogen oxides, while for copper, lead, nickel and uranium mining, the highest emission is from particulates< 2.5 mm. Also, the emission decreases after integrating solar collectors,

**TABLE 46** Life cycle assessment results based on the damage categories.

| Case | Human health | | | | Ecosystems quality | | | Climate change | | |
|---|---|---|---|---|---|---|---|---|---|---|
| | Total | Particulates, < 2.5 μm | Nitrogen oxides | Sulphur dioxide | Total | Aluminium | Nitrogen oxides | Total | Carbon dioxide, fossil | Dinitrogen monoxide |
| Bauxite base | 23 | | 19.1 | 0.28 | 0.78 | | 0.63 | 7.92 | 7.74 | 0.02 |
| Bauxite FPC | 21.6 | | 18.5 | 0.21 | 0.72 | | 0.61 | 6.02 | 5.88 | 0.02 |
| Bauxite ETC | 21.6 | | 18.5 | 0.21 | 0.72 | | 0.61 | 6.02 | 5.88 | 0.02 |
| Copper base | 4,027 | 1,290 | 630 | 2,220 | 1,470 | 1,009 | 20 | 605.6 | 588.21 | 6.63 |
| Copper FPC | 4,019 | 1,260 | 620 | 2,180 | 1,470 | 1,009 | 20 | 515 | 498 | 6.41 |
| Copper ETC | 4,018 | 1,260 | 610 | 2,180 | 1,470 | 1,009 | 20 | 513 | 496 | 6.4 |
| Ilmenite base | 2.89 | 9.04 | 14.1 | 3.48 | 4 | 0.64 | 0.46 | 30 | 29.5 | 0.08 |
| Ilmenite FPC | 2.3 | 7.95 | 12.5 | 0.75 | 3.7 | 0.18 | 0.41 | 23.1 | 22.7 | 0.07 |
| Ilmenite ETC | 2.3 | 7.87 | 12.4 | 0.72 | 3.6 | 0.23 | 0.41 | 22.9 | 22.6 | 0.07 |

| | | | | | | | | | | |
|---|---|---|---|---|---|---|---|---|---|---|
| Rutile base | 150 | 46 | 73 | 0.018 | 21 | 3.32 | 2.4 | 155.9 | 153.47 | 0.41 |
| Rutile FPC | 121 | 41 | 64.8 | | 19 | | 2.1 | 120 | 118 | 0.37 |
| Rutile ETC | 119 | 40.9 | 64.6 | 3.72 | 19 | 0.82 | 2.1 | 119 | 117 | 0.36 |
| Lead base | 2,630 | 90 | 16 | 0.28 | 265 | 181 | 5.34 | 238.7 | 233.7 | 1.33 |
| Lead FPC | 2,580 | 85 | 14.4 | | 261 | 176 | 4.78 | 178 | 173 | 1.25 |
| Lead ETC | 2,580 | 85 | 14.4 | | 261 | 176 | 4.78 | 177 | 172 | 1.24 |
| Nickel base | 13,650 | 1,840 | 92.2 | 10,630 | 2,270 | 1,480 | 30 | 1,215 | 1,188 | 10 |
| Nickel FPC | 13,500 | 1,780 | 89.2 | 10,500 | 2,270 | 1,490 | 29.6 | 962 | 934 | 9.62 |
| Nickel ETC | 13,500 | 1,770 | 89.1 | 10,500 | 2,270 | 1,490 | 29.6 | 956 | 929 | 9.6 |
| Uranium base | 251,000 | 82,170 | 11,310 | 5,300 | 89,790 | 71,690 | 370 | 10,200 | 9,753 | 7,639 |
| Uranium FPC | 249,000 | 81,300 | 11,000 | 4,350 | 89,790 | 71,700 | 364 | 8,050 | 7,790 | 7,050 |
| Uranium ETC | 249,000 | 81,300 | 11,000 | 4,350 | 89,790 | 71,700 | 364 | 8,010 | 7,750 | 7,040 |

specifically the particulates emission decreases more rather than for nitrogen oxides and sulphur oxides. For ecosystems quality category, the major responsible emissions are aluminium and nitrogen oxides. Other than for bauxite mining, all other mining has the highest contribution of emission in the air. The emission greatly reduces after integrating FPC and ETC while both showed the same performance to reduce the emissions. However, for copper and nickel mining, the environmental emission does not reduce after solar process heat integration. For the climate change category, the harmful emission is carbon dioxide fossil and dinitrogen monoxides, while carbon dioxide fossil has the greatest contribution. The results show that there is little change in dinitrogen monoxide emissions, while replacing fossil fuels using solar heat greatly reduces the carbon dioxide fossil emissions. In summary, the solar process heat integration helps to enhance sustainability under different damage categories like human health, ecosystems and climate change. The greatest emission reduction is evident for reducing the nitrogen oxides emission and carbon dioxide and fossil emissions.

## Discussion

In the carcinogen category, for aluminium, copper, nickel and uranium ore mining, the integration scenario of FPC and ETC gives better results with reduced environmental impact, and both are quite similar. For ilmenite, rutile and lead ore mining, the impact increases after solar process heat integration, while FPC is much harmful to the environment. In the noncarcinogen category, solar integration in aluminium and lead ore mining shows environment-friendly results. For nickel and copper ore mining processes, solar energy integration does not have any impact; the result remains the same as before. For ilmenite and rutile ore mining, base case and FPC integration show the same results, while ETC gives better performance. For uranium ore mining, FPC increases the environmental burden, while ETC integration has no impact.

In the respiratory inorganics and respiratory organics categories, it shows quite a stable outcome for both integration scenario of solar collectors. For all the mining process like aluminium, copper, ilmenite, rutile, lead and nickel, the environmental impact reduces from the base case through process heat integration, while both collectors have the same results. For nickel ore mining, the solar process heat integration does not have any impacts. The ozone layer depletion category and the process heat integration scenarios show better performance in case of copper, ilmenite, rutile, lead, nickel and uranium ore mining. For aluminium ore mining, the impact increases after solar process heat integration. Ionising radiation is an exceptional impact category, whereas solar process heat integration gives a negative performance while increasing the environmental burdens for aluminium, copper, ilmenite, rutile, nickel and lead ore mining. However, for uranium ore mining, there is no change in impact after process heat integration.

The impact categories based on acidification are terrestrial acidification and aquatic acidification. The ecotoxicity categories are aquatic ecotoxicity and terrestrial ecotoxicity. Both are assessed separately here and give the same kind of results. For all seven types of metal ore mining processes, the impact reduces after solar process heat integration, while ETC gives better performance. In the aquatic eutrophication, land occupation categories and variable results for each of the ore mining processes are evident here. For aluminium ore mining, the impact reduces after solar process heat integration, while ETC has better performance. For ilmenite, rutile, lead and uranium mining processes which are discussed here, the impact increases after solar process heat integration, while ETC has better performance. For copper and nickel ore mining, there is no change in the impact of this category due to solar process heat integration.

For the global warming categories, the results showed that for aluminium, ilmenite, nickel, lead and uranium ore mining the impact reduces after solar process heat integration, while both FPC and ETC give the same performance. For copper and rutile ore mining, the impact decreases more with the integration of ETC. From the LCA analysis, it is evident that solar process heat integration does not have much impact on the nonrenewable energy category. The analysis results presented for the damage-based categories also indicate that nitrogen oxides and carbon dioxide emissions greatly reduce after solar process heat integration, which would be beneficial for enhancing sustainability.

For most of the categories, this theory is proved to enhance environmental sustainability; however, for the land use and ionising radiation categories, using solar collector technologies does not bring any significant improvement as compared to using fossil fuels. The section-wise analysis results in Chapter 4 showed that solar process heating systems are much effective to reduce environmental burdens associated with the quality and resources of the ecosystem. From the discussion presented below, it is evidenced that solar industrial process heating systems can reduce the environmental impacts by at least 10% in all the after replacing fossil fuel technologies with solar collectors considered in this study. The reason behind this environmental emission reduction is due to reduced carbon dioxide, nitrogen oxides and particulate matter emission reduction due to the fossil fuel replacement by the renewable sources of process heat generation.

Fig. 28 shows the comparison among the three process heat generation systems based on environmental impact categories in aluminium/bauxite mining process. This figure shows how much the impact reduces after integrating solar process heating systems. From the diagram, it is evident that while bauxite mining has the highest impact on human health and climate

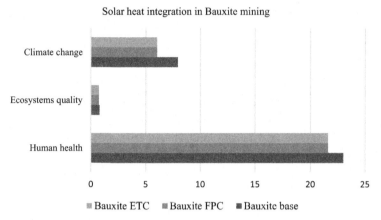

**FIGURE 28** Implications to theory from solar process heat integration in aluminium mining.

change, impact greatly reduces after integrating flat plate/ETC technologies, while both perform at a similar scale.

Fig. 29 shows the process of heat integration potential in copper ore mining industries. Significant impact reduction potential is apparent in the case of climate change, with little impact on human health and ecosystems quality.

Fig. 30 shows the process of heat integration potential in ilmenite ore mining industries. From the results presented in Fig. 30, it is obvious that environmental impacts can be reduced in the categories of climate change, with little reduction in ecosystems quality and human health.

Fig. 31 shows the process of heat integration potential in rutile ore mining industries. From the results presented in Fig. 31, it is apparent that

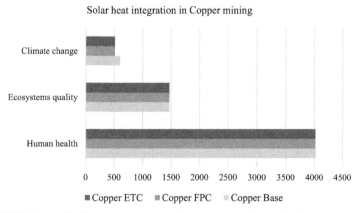

**FIGURE 29** Implications to theory from solar process heat integration in copper mining.

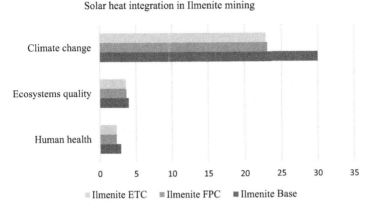

**FIGURE 30**    Implications to theory from solar process heat integration in ilmenite mining.

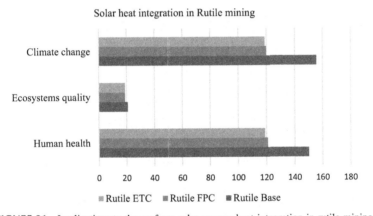

**FIGURE 31**    Implications to theory from solar process heat integration in rutile mining.

environmental impacts can be reduced in the categories of climate change and human health. Impacts can also be reduced in the categories of ecosystems quality.

Fig. 32 shows the process of heat integration potential in lead ore mining industries. From the results presented in Fig. 32, it is clear that impacts reduce for all of the damage categories, while both of the collectors have similar performance.

Fig. 33 shows the process of heat integration potential in lead ore mining industries. From the results presented in Fig. 33, solar heat integration can reduce impacts on human health and climate change, with no impact on the emissions for ecosystems quality.

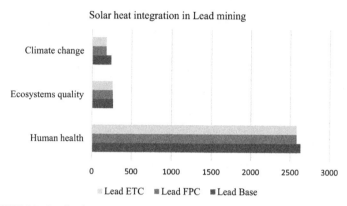

**FIGURE 32** Implications to theory from solar process heat integration in lead mining.

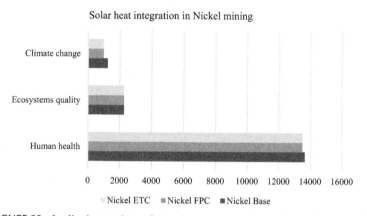

**FIGURE 33** Implications to theory from solar process heat integration in nickel mining.

Fig. 34 shows the process of heat integration potential in uranium ore mining industries. The graph shows that environmental burdens can be reduced in the categories of climate change and human health with almost no change in ecosystems quality. However, both collectors have similar performance.

In respect of cleaner production/sustainability, this chapter aims to identify the potentialities of solar industrial process heat integration in mining industries. The theory implied from this chapter is that ETC integration would be beneficial for cleaner production/sustainability rather than the existing practice of process heat generation using fossil fuels. The key factors which are considered for the foundation of the theory related to solar process heat integration potential are the environmental impact categories analysed through LCA technologies. Analysis results implied that the environmental impacts

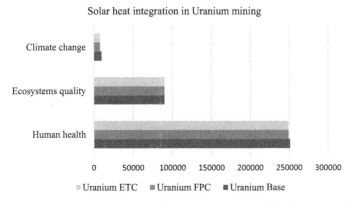

**FIGURE 34** Implications to theory from solar process heat integration in uranium mining.

under the 15 major categories (midpoint indicator-based categories) reduce significantly after integrating evacuated tube solar collector systems for process heat generation. Using the process heating system with FPC can also reduce the environmental burdens to a great extent, but not as much as the ETCs.

## Conclusion

This chapter aimed to analyse the environmental impact reduction possibilities after solar industrial process heating systems integration in the selected mining and mineral processing industries. The selected industries are aluminium, copper, ilmenite, rutile, lead, nickel and uranium. The solar collectors considered here were FPC and ETC. The results of the analysis presented in this chapter show that uranium, ilmenite, rutile and aluminium ore mining will have better results in terms of reducing impacts after integrating solar industrial process heating systems. This study also reveals that ETCs are much environment-friendly than FPCs, reducing environmental burdens significantly. The solar process heat integration scenario shows that the integration of renewable energy resources could notably reduce impacts on global warming, carcinogens, respiratory inorganics, aquatic and terrestrial ecotoxicity and acidification. Overall, solar industrial process heating system integration in ore mining processes has immense potential to reduce environmental impacts from ore mining. This chapter pointed out that among the considered mining processes, the noteworthy environmental burden is associated with copper, nickel and uranium mining, which requires the focus of attention. Future research should focus on these mining processes with prime importance. In future, this research can further progress based on the PinCH analysis to assess the feasibility of integration in the mining processes to identify the integration point. This study can also be extended to design the appropriate solar thermal system using evacuated tube solar collector, which finds out to be more beneficial based on sustainability from the present research.

# References

Adiansyah, J.S., Haque, N., Rosano, M., Biswas, W., 2017. Application of a life cycle assessment to compare environmental performance in coal mine tailings management. J. Environ. Manag. 199, 181−191. https://doi.org/10.1016/j.jenvman.2017.05.050.

Althaus, H., Chudacoff, M., Hischier, R., Jungbluth, N., Osses, M., Primas, A., Hellweg, S., 2007. Life Cycle Inventories of Chemicals. Ecoinvent Data v2.0. Swiss Centre for Life Cycle Inventories, pp. 1−957.

Althaus, H.J., Classen, M., 2005. Life cycle inventories of metals and methodological aspects of inventorying material resources in ecoinvent. Int. J. Life Cycle Assess. 10, 43−49. https://doi.org/10.1065/lca2004.11.181.5.

Bartzas, G., Komnitsas, K., 2015. Life cycle assessment of ferronickel production in Greece. Resour. Conserv. Recycl. 105, 113−122. https://doi.org/10.1016/j.resconrec.2015.10.016.

Beath, A.C., 2012. Industrial energy usage in Australia and the potential for implementation of solar thermal heat and power. Energy 43, 261−272. https://doi.org/10.1016/j.energy.2012.04.031.

Chen, W., Geng, Y., Hong, J., Dong, H., Cui, X., Sun, M., Zhang, Q., 2018. Life cycle assessment of gold production in China. J. Clean. Prod. 179, 143−150. https://doi.org/10.1016/j.jclepro.2018.01.114.

Chen, W., Geng, Y., Tian, X., Zhong, S., Gao, C., You, W., Li, S., 2019. Emergy-based environmental accounting of gold ingot production in China. Resour. Conserv. Recycl. 143, 60−67. https://doi.org/10.1016/j.resconrec.2018.12.021.

Doka, G., 2011. Life Cycle Inventory of Generic Uranium In-Situ Leaching.

Duffie, J.A., Beckman, W.A., Worek, W.M., 2003. Solar engineering of thermal processes. Fourth ed. J. Sol. Energy Eng. https://doi.org/10.1115/1.2930068.

EPA, 2018. The Emissions & Generation Resource Integrated Database (eGRID) Technical Support Document, vol. 106. US Environ. Prot. Agency.

European Solar Thermal Industry Federation, 2006. Solar industrial process heat. State of the art. Key Issues Renew. Heat Eur. 1−15.

Farjana, S.H., Huda, N., Mahmud, M.A.P., 2018a. Life-cycle environmental impact assessment of mineral industries. In: IOP Conference Series: Materials Science and Engineering, 351. IOP Publishing, pp. 1−7, 012016.

Farjana, S.H., Huda, N., Mahmud, M.A.P., 2019a. Life cycle analysis of copper-gold-lead-silver-zinc beneficiation process. Sci. Total Environ. 659, 41−52. https://doi.org/10.1016/j.scitotenv.2018.12.318.

Farjana, S.H., Huda, N., Mahmud, M.A.P., 2019b. Life cycle assessment of cobalt extraction process. J. Sustain. Min. 18 (3), 150−161. https://doi.org/10.1016/j.jsm.2019.03.002.

Farjana, S.H., Huda, N., Mahmud, M.A.P., Lang, C., 2018b. Comparative life-cycle assessment of uranium extraction processes in Australia. J. Clean. Prod. 202, 666−683. https://doi.org/10.1016/j.jclepro.2018.08.105.

Farjana, S.H., Huda, N., Mahmud, M.A.P., Lang, C., 2018c. Towards sustainable $TiO_2$ production: an investigation of environmental impacts of ilmenite and rutile processing routes in Australia. J. Clean. Prod. 196, 1016−1025. https://doi.org/10.1016/j.jclepro.2018.06.156.

Farjana, S.H., Huda, N., Mahmud, M.A.P., Lang, C., 2019c. Impact analysis of gold-silver refining processes through life-cycle assessment. J. Clean. Prod. https://doi.org/10.1016/j.jclepro.2019.04.166.

Farjana, S.H., Huda, N., Mahmud, M.A.P., Saidur, R., 2018d. Solar industrial process heating systems in operation — current SHIP plants and future prospects in Australia. Renew. Sustain. Energy Rev. 91 https://doi.org/10.1016/j.rser.2018.03.105.

Farjana, S.H., Huda, N., Mahmud, M.A.P., Saidur, R., 2018e. Solar process heat in industrial systems — a global review. Renew. Sustain. Energy Rev. 82 https://doi.org/10.1016/j.rser.2017.08.065.

Farjana, S.H., Huda, N., Parvez Mahmud, M.A., 2018f. Environmental impact assessment of european non-ferro mining industries through life-cycle assessment. In: IOP Conference Series: Earth and Environmental Science, vol. 154. IOP Publishing, p. 012019. https://doi.org/10.1088/1755-1315/154/1/012019.

Farjana, S.H., Huda, N., Parvez Mahmud, M.A., Saidur, R., 2019d. A review on impact of mining and mineral processing industries through life cycle assessment. J. Clean. Prod. 231, 1200—1217. https://doi.org/10.1016/j.jclepro.2019.05.264.

Fogler, S., Timmons, D., 1998. An Overview of the ISO 14040 Life Cycle Assessment Approach and an Industrial Case Study. Argentum V.

Goedkoop, M., Oele, M., Vieira, M., Leijting, J., Ponsioen, T., Meijer, E., 2014. SimaPro Tutorial, vol. 89. https://doi.org/10.1142/S0218625X03005293.

Guimarães da Silva, M., Costa Muniz, A.R., Hoffmann, R., Luz Lisbôa, A.C., 2018. Impact of greenhouse gases on surface coal mining in Brazil. J. Clean. Prod. 193, 206—216. https://doi.org/10.1016/j.jclepro.2018.05.076.

Haque, N., Norgate, T., 2014. The greenhouse gas footprint of in-situ leaching of uranium, gold and copper in Australia. J. Clean. Prod. 84, 382—390. https://doi.org/10.1016/j.jclepro.2013.09.033.

Hess, S., Oliva, A., Stryi-hipp, G., 2011. Solar Process Heat — System Design for Selected Low-Temperature Applications in the Industry Solar Process Heat — System Design for Selected Low-Temperature Applications in the Industry. https://doi.org/10.18086/swc.2011.23.07.

Hischier, R., Weidema, B., Althaus, H.-J., Bauer, C., Doka, G., Dones, R., Frischknecht, R., Hellweg, S., Humbert, S., Jungbluth, N., Köllner, T., Loerincik, Y., Margni, M., Nemecek, T., 2010. Ecoinvent Rep. No. 3. Implementation of Life Cycle Impact Assessment Methods Data V2.2 (2010), vol. 176.

Humbert, S., Margni, M., Jolliet, O., PRe, Various authors, 2012. Quantis-sustainability counts. Impact 2002+: User Guide, vol. 21, p. 36. https://doi.org/10.1007/s10973-014-4166-8.

IEA-ETSAP, IRENA, 2015. Solar Heat for Industrial Processes -Technology Brief, vol. 37.

Khoo, J.Z., Haque, N., Woodbridge, G., McDonald, R., Bhattacharya, S., 2017. A life cycle assessment of a new laterite processing technology. J. Clean. Prod. 142, 1765—1777. https://doi.org/10.1016/j.jclepro.2016.11.111.

Lehtinen, H., Saarentaus, A., Rouhiainen, J., Pits, M., Azapagic, A., 2011. A review of LCA methods and tools and their suitability for SMEs. Eco Innov. Biochem. 24 https://doi.org/10.1017/CBO9781107415324.004.

Lillo, I., Pérez, E., Moreno, S., Silva, M., 2017. Process heat generation potential from solar concentration technologies in Latin America: the case of Argentina. Energies 10. https://doi.org/10.3390/en10030383.

Mahmud, M., Huda, N., Farjana, S., Lang, C., Mahmud, M.A.P., Huda, N., Farjana, S.H., Lang, C., 2018a. Environmental impacts of solar-photovoltaic and solar-thermal systems with life-cycle assessment. Energies 11, 2346. https://doi.org/10.3390/EN11092346.

Mahmud, M.A.P., Huda, N., Farjana, S.H., Lang, C., 2019. A strategic impact assessment of hydropower plants in alpine and non-alpine areas of Europe. Appl. Energy 250, 198−214. https://doi.org/10.1016/j.apenergy.2019.05.007.

Mahmud, M.A.P., Huda, N., Farjana, S.H., Lang, C., 2018b. Environmental sustainability assessment of hydropower plant in Europe using life cycle assessment. In: IOP Conference Series: Materials Science and Engineering. https://doi.org/10.1088/1757-899X/351/1/012006.

Memary, R., Giurco, D., Mudd, G., Mason, L., 2012. Life cycle assessment: a time-series analysis of copper. J. Clean. Prod. 33, 97−108. https://doi.org/10.1016/j.jclepro.2012.04.025.

Norgate, T., Haque, N., 2010. Energy and greenhouse gas impacts of mining and mineral processing operations. J. Clean. Prod. 18, 266−274. https://doi.org/10.1016/j.jclepro.2009.09.020.

Norgate, T.E., Rankin, W.J., 2000. Life cycle assessment of copper and nickel production. In: Int. Conf. Miner. Process. Extr. Metall. (MINPREX), Melbourne, 11−14 Sept. 2000, pp. 133−138.

Northey, S., Haque, N., Mudd, G., 2013. Using sustainability reporting to assess the environmental footprint of copper mining. J. Clean. Prod. 40, 118−128. https://doi.org/10.1016/j.jclepro.2012.09.027.

Norton, B., 2012. Industrial and agricultural applications of solar heat. Compr. Renew. Energy 3, 567−594. https://doi.org/10.1016/B978-0-08-087872-0.00317-6.

Nunez, P., Jones, S., 2016. Cradle to gate: life cycle impact of primary aluminium production. Int. J. Life Cycle Assess. 21, 1594−1604. https://doi.org/10.1007/s11367-015-1003-7.

PRé, 2018. SimaPro Database Manual Methods Library Colophon Title: SimaPro Database Manual Methods Library. https://doi.org/10.1017/CBO9781107415324.004.

International Renewable Energy Agency, 2014. Renewable Energy Options for the Industry Sector: Global and Regional Potential until 2030.

Ruan, S., Unluer, C., 2016. Comparative life cycle assessment of reactive MgO and Portland cement production. J. Clean. Prod. 137, 258−273. https://doi.org/10.1016/j.jclepro.2016.07.071.

Sharma, A.K., Sharma, C., Mullick, S.C., Kandpal, T.C., 2017. Solar industrial process heating: a review. Renew. Sustain. Energy Rev. 78, 124−137. https://doi.org/10.1016/j.rser.2017.04.079.

Tan, R.B.H., Khoo, H.H., 2005. An LCA study of a primary aluminum supply chain. J. Clean. Prod. 13, 607−618. https://doi.org/10.1016/j.jclepro.2003.12.022.

# List of Figures

# List of Tables

# Index

*Note*: 'Page numbers followed by "*f*" indicate figures and "*t*" indicate tables'.

Printed in the United States
By Bookmasters